S0-FIV-225

ENZYME KINETICS

A LEARNING PROGRAM FOR STUDENTS OF THE BIOLOGICAL AND MEDICAL SCIENCES

SECOND EDITION

HALVOR N. CHRISTENSEN, Ph.D.
Professor
Department of Biological Chemistry
The University of Michigan
Ann Arbor, Michigan

and

GRAHAM A. PALMER, Ph.D.
Professor of Biochemistry
Rice University
Houston, Texas

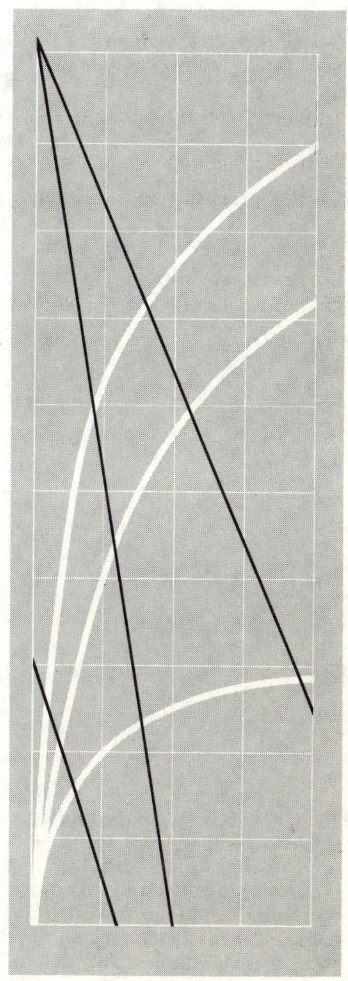

W. B. SAUNDERS COMPANY
Philadelphia, London, Toronto
1974

W. B. Saunders Company: West Washington Square
Philadelphia, PA 19105

12 Dyott Street
London, WC1A 1DB

833 Oxford Street
Toronto, Ontario M8Z 5T9, Canada

Listed here is the latest translated edition of this book together with the language of the translation and the publisher.

Italian (1st Edition) — Angeli Editore,
Milan, Italy

French (1st Edition) — Edicience S.A.
Paris, France

German (1st Edition) — Verlag Chemie
Weinheim, Germany

Enzyme Kinetics ISBN 0-7216-2591-6

Last digit is the print number: 9 8 7 6 5 4 3 2 1

Preface to the Second Edition

We are pleased that the response to the publication of *Enzyme Kinetics* has been enthusiastic enough to necessitate this revision of the book.

The most conspicuous changes in this second edition are insertions as follows:

1. A sequence on the allosteric phenomenon, Items 126 to 136, actually incorporated first into the French edition of 1971.
2. A section on binding studies, Items 141 to 159; also a prologue to that subject, in Items 10 to 12.
3. A short section to consider implications of kinetics for practical enzyme assays, Items 172 to 183.
4. Two items (97 and 98) to call the students' attention to two-substrate kinetics, without attempting to teach the subject. Among the new items, only these two have been given stars to commend that no user omit them.

Our own experience reinforces the helpful recommendations received from colleagues around the world that these additions be made. In doing so, we have tried to keep the program a short primer (a "modular" program, one that can be inserted easily into various instructional sequences) rather than letting it grow beyond such utility.

Why did we place the study of the binding of small molecules of proteins here rather than, as would seem logical, in a companion program on thermodynamics?* The equations that de-

* H. N. Christensen and R. A. Cellarius: *Introduction to Bioenergetics: Thermodynamics for the Biologist*. Philadelphia, W. B. Saunders Co., 1972.

scribe the extent of binding, even though they concern the state of equilibrium, are inherently the same as those that describe the kinetics of a catalytic process, since the velocity of the latter is taken to depend on the extent to which the substrate is bound at the catalytic site. Furthermore, the graphic methods used for the two purposes are parallel. Therefore we exploit here, for those who elect, a significant pedagogic advantage.

We must warn the reader not to regard unstarred items as less important than starred items. The items are unstarred merely because there may be categories of students who might not include the contained information among their immediate goals. For your goals, a given sequence of *unstarred* items may instead be especially important. Accordingly, our starring of a bare minimum of items still offering continuity should not prejudice the reader against any part of the subject.

We continue to welcome suggestions for improvements.

HALVOR N. CHRISTENSEN
GRAHAM A. PALMER

Preface to the First Edition

The gratifying acceptance of the two "modular" programed texts by the first author, *pH and Dissociation* and *Body Fluids and The Acid-Base Balance,* has led him during the last several years to try to persuade one or another enzyme kineticist to program that subject. The subject of enzyme kinetics seemed to be an ideal one for the programed treatment: It is relatively stable; it is demanding; and there is widespread dissatisfaction with the results obtained in teaching it. The upshot of that campaign was that when the second author became available as a collaborator, it was *ourselves* we persuaded. The attempt, now carried through three stages of experimental use, has fully confirmed our confidence that the subject would be particularly amenable to programed treatment.

Some explanations to our colleagues may be in order as to the scope, sequence, and emphasis of the program. The scope of the ideas presented appears to us to correspond to the treatment in the several leading textbooks of biochemistry, whether for science students or for medical students. This is a treatment for which an hour's lecture might ordinarily suffice. The objectives correspond to what we have found appropriate for those two groups of students, perhaps omitting unstarred items for the medical student. A possible impression that the program presents a significantly more intense treatment than that will not be confirmed on examination, we believe, if the objectives implied by the usual pithy textual treatments are taken at face value. For the student ready for programed assistance in understanding the subject, these hundred and forty-seven items or steps represent what we find to approximate the self-directed explanation needed.

As to the sequence by which our subject is developed, our tests, aided by suggestions from colleagues, showed us the advantage of treating the *general* case before the *special*, even when that sequence inverts the historic one. Thus the derivation of the Michaelis-Menten equation under the steady state approximation of Briggs and Haldane precedes the less frequently applicable derivation under the equilibrium assumption. Similarly, the advantage became evident of considering first the case of the inhibitory analog that can also serve as a substrate, and then afterwards the more restricted case in which no attack can occur. That sequence, incidentally, is particularly helpful to the student who will apply kinetics to the study of transport, for which the more limited conception of the competitive analog is usually not applicable. (What you teach your student as enzyme kinetics may serve him for quite a different biological or pharmacological phenomenon.) Besides, competitively reactive analogs are by no means rare in enzymology.

The learner is strongly advised actually to complete each item as he goes along, whether by writing in a word or by completing a graph, rather than merely to imagine its completion. He must expect to repeat sections which the review material or the terminal examination shows he has not adequately understood. Instructors should be aware of a strange phenomenon whereby the student may consider acquisition of a learning program nearly equivalent to mastery of the subject.

Finally we want to thank numerous colleagues who have

made valuable suggestions for the improvement of this program at one or another of the experimental stages. Experimental use has been made by the following, outside this University: Dr. R. H. DeMeio, Jefferson Medical College; Dr. John Lyon, Emory University; Dr. Carl Vestling, University of Iowa; and Dr. Lars Broman and Dr. Göran Pettersson, University of Uppsala. We are also indebted to Professor Emeritus G. E. Briggs of Cambridge University and to the *Biochemical Journal* for consenting to our reproducing the short, historic paper of Briggs and Haldane.

Halvor N. Christensen
Graham A. Palmer

Ann Arbor, Michigan

Contents

APPENDICES

INTRODUCTION

1

* This program will consider four factors affecting the velocity of enzymatic reactions:

A. The concentration of the substance undergoing the enzymatic reaction (namely the *substrate*)

B. The concentration of an inhibitor of the reaction

C. The temperature

D. The pH

The study of the first of these is important on several counts:

1. For what it tells us about how enzymes act in general

2. To describe the characteristics of a given enzyme

3. To show what conditions are necessary for enzymatic analysis, whether to determine the concentration of an enzyme or to use an enzyme to determine the concentration of a substrate

2

* If we were to use the rate of an enzymatic reaction to determine the concentration of an enzyme, for example in the blood plasma, we should want the rate to be limited by the amount of the enzyme present. Hence the substrate should be present in | stoichiometric amounts | generous excess |. Similarly any cofactor must be present in | excess | limiting concentration |.

3

* To study the effect of substrate concentration on the velocity, we are interested instead in the range of substrate concentrations where the rate increases with a rising level of the substrate. For this purpose we | must | ordinarily would not | use the substrate in generous excess, as we did in Item 2 for assaying enzyme activity.

1

2

generous excess

excess

3

ordinarily would not

ORIGIN OF LIMIT ON RATE SET BY AMOUNT OF ENZYME PRESENT; THE *ES* COMPLEX

4

* The figure shows how in many chemical reactions the rate increases with the concentration of the reactant. The straightness of the line shows that no factor other than the concentration of the reactant begins to limit the rate within the concentration range studied. Suppose the reaction is enzymatic, and at high substrate concentrations the amount of enzyme available becomes limiting. The rate will then be expected to change somewhat as indicated by curve | *A* | *B* | *C* |.

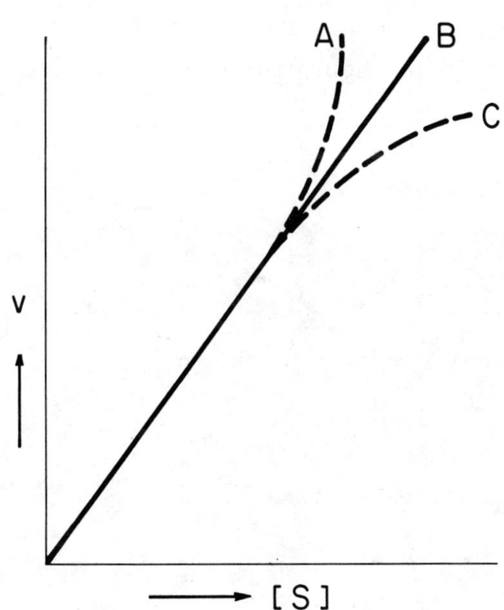

5 4

* The following figure shows with an arbitrarily selected example how the velocity of enzymic reactions characteristically increases with substrate concentration. (The curve has been interrupted to permit examination over a wide range of concentrations.) At first, at low substrate concentrations, the rate does rise almost linearly with concentration. But at higher concentrations the increase gradually lessens, until finally the reaction can hardly be accelerated further by increasing substrate concentration. Because the rate can no longer be made to rise, we say that the substrate has *saturated* the enzymatic process. The rate observed under the latter conditions is called the *maximal velocity,* indicated by V_{max} or sometimes simply a capital V.

C

(Note that this definition does not say that the V_{max} is the highest rate that can be reached under any conditions whatever, e.g., by a higher temperature.)

In the figure the value of V_{max} appears to be about __________________.

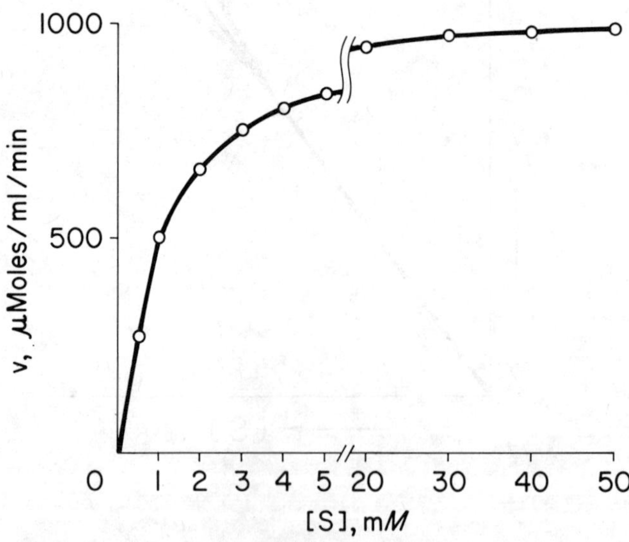

5

1000 micromoles per ml · min (micromoles ml^{-1} min^{-1})

6

* The figure in Item 5 also brings out the point that the maximal velocity can be approximated with good accuracy [by using only modest | only by using relatively high] concentrations of substrate, because this rate is approached asymptotically as the substrate concentration is made larger and larger.

6

only by using relatively high

(The terms, regrettably, are quite subjective.)

7

* Now, what is the nature of the limitation that causes the rate to stop rising steadily with substrate concentration in the figure in Item 5? That is, why does the substrate tend to ____________ the enzymatic reaction? This behavior shows that we have entered a range of substrate concentrations in which the amount of the enzyme available begins to be rate-limiting. The origin of this limitation touches on the question of what the enzyme is doing to cause the chemical reaction to take place. The hypothesis of Henri (1902), subsequently adopted by Michaelis and Menten (1913), held that the enzyme, E, forms with the substrate, S, an intermediate *complex*, ES, from which the product is formed:

$$E + S \rightleftharpoons ES$$

We have written this equation to show that it is [reversible | irreversible].

8

7

* Subsequently, the formation of enzyme-substrate complexes has been demonstrated experimentally on several occasions. Usually the formation of a new substance when E and S are mixed has been signaled by a new absorption band (or even a new color, when the absorption falls in the visual range), one that cannot be attributed to the product. Accordingly we must regard the formation of the ES complex as (*check the correct answer*):

( ) a convenient but unsupported hypothesis

( ) a reality

saturate

reversible

9

8

* Indeed, enzymologists have learned by a number of procedures to recognize a particular region of or cleft in the surface of the enzyme molecule called the *active site* with which the ES complex is formed, and to determine which amino acid sidechains of the protein structure contribute to the specificity and firmness of the binding at this __________ __________, and which amino acid sidechains determine what catalytic effect occurs.

a reality

9

active site

10

Although we have begun here by emphasizing the binding of small substrate molecules by enzymes, it is important to note that most, if not all, globular proteins show the capacity for selectively binding small molecules. Some other cases in which these binding properties serve biologically useful purposes may be listed:

Carrier proteins in circulating fluids; for example, hemoglobin, transferrin.

Membrane carrier proteins (as considered later in this program).

Immune proteins, in particular antibodies.

Signalling proteins, with receptor sites at cell surfaces; for example, chemotactic receptors, and sites activating adenylate cyclase.

The molecules bound at these sites appear usually not to be modified by the reversible binding event, and analogs that cannot be modified often mimic the effects of the substance ordinarily bound. For lack of evidence to the contrary, we may therefore provisionally suppose that the proteins on this list [resemble | do not resemble] enzymes in catalyzing a chemical transformation of the bound molecule.

11

Some of these sites are recognized to serve for regulation of cellular processes. Indeed, enzymes themselves have binding sites other than those at which binding leads to chemical transformation. Some of these other sites have been shown to serve as *modifier sites*. That is, when the site is occupied, the action of the enzyme is accelerated or slowed, as we shall consider later in this program.

Did the preceding paragraph imply that enzymes regularly have two different catalytic sites?

10

do not resemble

12

Still other binding sites can be detected on globular proteins which have no known biologic function. We may take the presence of such sites to be an inherent consequence of globular protein structure.

These statements | imply | do not imply, however, | that all globular protein act as enzymes. We will now proceed to describe how the binding property is conventionally evaluated for the purposes of enzyme kinetics. Near the end of the program we will note some of the other useful ways in which the binding property itself is studied.

11

No (One is catalytic; others, if present, were described as modifier sites.)

12

do not imply, however,

13

ES

P

13

* The enzyme-substrate complex (which is indicated in the above equation by the symbol ______) may react either to regenerate the reactants or, as shown here

$$ES \rightleftharpoons P + E$$

to release the product, ______, and the original unmodified enzyme, E.

14

* We will now permit ourselves to consider this second reaction as irreversible, by agreeing to make our observations at a negligible concentration of P, so that no significant reversal can occur. Accordingly, for the process

$$E + S \rightleftharpoons ES \longrightarrow P + E$$

we must make our observations:

( ) shortly after adding enzyme to the substrate solution

( ) after the reaction has proceeded for a long time

15

* Ordinarily one can select a time interval short enough so that the amount of P formed is still so small as to preclude significant back reaction. Enzymatic reaction rates measured under these conditions are called *initial-rate* measurements. In more sophisticated analyses the back-reaction is also taken into consideration. Unless this is done one must certainly observe the rate before a significant amount of ______ has had time to accumulate. (One could of course study instead the reverse reaction of $E + P$, provided in that case that he makes his observations before a significant amount of ______ accumulates.)

14

shortly after adding enzyme to the substrate solution

16

* From examining the chemical equation

$$E + S \rightleftharpoons \_\_\_\_\_\_ \longrightarrow P + \_\_\_\_\_\_$$

we can see that the enzyme is tied up in the reaction for a finite interval of time, thereby limiting the rise in initial reaction rate with substrate concentration, as shown in the figure in Item 5.

15

P*

S

*Not ES. One cannot begin to observe the typical progress of the forward reaction until ES has formed to the steady state level. A glance ahead to the figure in Item 36 will show that *too short* a time interval will not serve.

'E CONSTANTS FOR FIRST- AND :OND-ORDER REACTIONS; EQUILIBRIUM NSTANTS

[o understand, however, why the curve of the igure in Item 5 shows the shape it does (a ectangular hyperbola) we must consider the asis of the change of reaction velocity with he concentrations of the reactants, that is, he *law of mass-action.*

Calculus provides us with a useful form or expressing *rate* (i.e., of how the change in ne variable relates to the change in another) hich is easily understood even if one has tudied almost no calculus. By this convention, or the chemical reaction

$$A + B \rightleftharpoons Q + R$$

ie symbol for the rate at which a reactant is consumed as a function of time is

$$-\frac{d[A]}{dt}$$

ne may read this as the "change in A for a ven small (strictly, infinitesimal) change in me." The negative sign shows that we are scussing the | appearance | disappearance | the reactant.

18

* Similarly the appearance of a product P is represented by

$-\frac{d[P]}{dt}$	$+\frac{d[P]}{dt}$

19

* In order to follow the over-all rate of a given enzyme reaction in a given solution, we need either a method to measure A so that we can see how much has been | formed | consumed |, or a method of measuring P so we can see how much has been ____________.

20

* The law of mass-action states, for a general case

$$A + B \rightleftharpoons Q + R$$

that the velocity of the forward reaction is directly proportional to the product of the concentration of the reactants, thus

$$v_f \propto [A] \cdot [B]$$

and that the rate of the reverse reaction is given by

$$v_r \quad \_\_\_\_\_\_\_\_\_\_\_\_$$

17

disappearance

18

$$+\frac{d[P]}{dt}$$

19

consumed

formed

20

$\propto [Q] \cdot [R]$

21

* These statements can be rewritten:

$$v_f = k_f \cdot [\_\_\_\_\_] \cdot [\_\_\_\_\_]$$

$$v_r = k_r \cdot \_\_\_\_\_\_\_\_$$

21

A, B

$[Q] \cdot [R]$

22

* The proportionality constants k_f and k_r are known as *rate constants.* To avoid confusion with certain other constants we will take pains to use lower-case k's for rate constants.

We will first consider a special case, where $A \longrightarrow Q + \ldots$, for which the forward rate, $-\frac{d[A]}{dt}$, is given by:

$$v_f = k_f [A]$$

Rearranging:

$$k_f = \_\_\_\_\_\_\_\_$$

The rate constant is called a *first-order* rate constant when the rate depends directly, as in this instance

( ) on the first power of the concentration of a single reactant

( ) on a higher power of the concentration of a single reactant

( ) on the concentration of more than one reactant

23

* For the case

$$A + B \longrightarrow Q + \ldots$$

the constant k_f is known instead as a ______ order rate constant. If we write $v_f = k_f[A] \cdot [B]$, we can calculate that $k_f =$ ____________. (If one molecule of A must react with a second molecule of A, thus

$$2A \longrightarrow Q + \ldots$$

the rate constant will also be a ____________ order constant.)

24

It is worth noting that *first-order* and *second order* rate constants do not have identical dimensions. Compare the two defining equations:

$$k = v/[A]$$

$$k = v/[A] \cdot [B]$$

$$(\text{or, } k = v/[A]^2)$$

Since v is given in the dimension, concentration/time, and $[A]$ and $[B]$ are given in concentration units, any first-order constant has the dimension

$$\frac{\textit{concentration/time}}{\textit{concentration}}$$

which simplifies to ____________, for example min^{-1} or sec^{-1}.

22

$\frac{v_f}{[A]}$

on the first power of the concentration of a single reactant

23

second

$\frac{v_f}{[A] \cdot [B]}$

second

24

time^{-1} (1/time)

25

From the second equation in Item 24, we see that a second-order rate constant will be given in

$$\frac{concentration/time}{concentration \cdot concentration}$$

Simplifying, we see that the constant has the dimension ________________.

25

concentration^{-1} × time^{-1}

26

If one rate constant is given as 800 per micromolar per second, and another is given simply as 600 per second, we may presume that the former is a __________-__________ rate constant and the latter is a __________-__________ rate constant.

26

second-order

first-order

27

Let us consider another simple reaction,

$$P + L \underset{k_{-1}}{\overset{k_1}{\rightleftharpoons}} PL$$

When the concentrations of the several components stop changing with time, we say that *equilibrium* has been reached. We then use the equilibrium concentrations reached (designated by the subscript e) to define an equilibrium constant,

$$K_e = \frac{[PL]_e}{[P]_e \times [L]_e}$$

which is the ratio of the product of the concentrations of all products (here only PL) to the product of the concentration of all ________________.

28

As written here, K_e is an *association constant* (K_a) and its dimensions are [$\underline{M}$ | $\underline{M}^{-1}$]. Typically, its magnitude will fall in the range 10^3 to 10^7.

Alternatively, we could have written the above reaction in the form,

$$PL \rightleftharpoons P + L$$

with

$$K_e = \frac{[P]_e \times [L]_e}{[PL]_e}$$

K_e is then called a dissociation constant, K_d. In this form it has the dimensions of [$\underline{M}$ | $\underline{M}^{-1}$], and its magnitude usually falls in the range [10^3 to 10^7] [10^{-3} to 10^{-7}]

29

Note the important relation between the equilibrium constants and the rate constants for a reaction: The state of equilibrium implies that the rate of conversion of P and L into PL (namely, k_1 $[P] \cdot [L]$ as in Item 27) is equal to the rate of the dissociation of PL into P and L (namely k_{-1} $[PL]$). That is,

$$k_1 [P]_e \cdot [L]_e = k_{-1} [PL]_e$$

hence

$$\frac{k_1}{k_{-1}} = \frac{[PL]_e}{[P]_e \times [L]_e} = K_a$$

or, by inverting, [ ] $= K_d$

(Show each term in the equation in inverted form.)

27

reactants

28

$\underline{M}^{-1}$

$\underline{M}$

10^{-3} to 10^{-7}

29

$$\frac{k_{-1}}{k_1} = \frac{[P]_e \times [L]_e}{[PL]_e}$$

DERIVATION OF MICHAELIS-MENTEN EQUATION UNDER STEADY-STATE APPROXIMATION

30

* Let us return then to the rate of enzymatic reactions. We will use the symbols $[E]$, $[S]$, $[ES]$ and $[P]$, respectively, for the concentrations of enzyme, the substrate, the ________ ____________ ____________ and the product.* To these four terms we will add $[E_t]$, the total concentration of enzyme, whether free or coupled. That is:

$$[E_t] = [E] + [\_\_\_\_\_\_]$$

30

enzyme-substrate complex

ES

31

* (We might suppose that a similar term, $[S_t]$, should be introduced for the total concentration of substrate initially present, whether free, enzyme-bound or, at a given point in time, already converted to the product. Two conditions, however, make it possible to use $[S]$ as if it had interchangeably the meaning $[S]$ or $[S_t]$. First, the amount of substrate present generally will be very much larger than the amount of enzyme; hence | a large | only a small | proportion of the substrate will be enzyme-bound. Second, as mentioned in Item 11, if one studies the initial velocity of an enzymatic reaction, the proportion of S converted to P during an assay will be very | large | small |.)

*Another convention uses e for total enzyme concentration and x for the concentration of ES. The value of $[E]$ is therefore $e - x$. Between the two usual conventions for identifying rate constants, we will choose (beginning in Item 32) a standard one by which a negative subscript identifies a rate constant for a reaction proceeding from right to left.

32

* From the equation in Item 11

$$E + S \underset{k_{-1}}{\overset{k_1}{\rightleftharpoons}} ES \xrightarrow{k_2} P + E$$

we can see that the rate of formation of P will depend on the concentration of ES; that is:

$$v = k_2[\_\_\_\_\_\_]$$

33

* Furthermore we can apply the law of mass action to describe how the concentration of the enzyme-substrate complex will change with time. The rate at which ES will be *formed* will equal $k_1[E][S]$; the rate at which it will be *converted back* to E and S will equal $k_{-1}[ES]$; and the rate at which it will be *converted to product* plus regenerated enzyme will be $k_2[ES]$. Hence we may write:

$$\frac{d[ES]}{dt} = k_1[E][S] - k_{-1}[ES] \_\_\_\_\_\_\_\_$$

(*Please complete the equation.*)

34

* Recall that $[E_t] = [ES] + [E]$; hence $[E] = [E_t] - [ES]$. If we substitute that value for $[E]$ into the right-hand side of the equation, we obtain:

$$\frac{d[ES]}{dt} = \_\_\_\_\_\_\_\_\_\_\_\_\_\_\_\_$$

31

only a small

small

32

ES

33

$-k_2[ES]$

34

$k_1([E_t]-[ES])[S]-$
$k_{-1}[ES]-k_2[ES]$

35

* Our objective will be to manipulate these relationships to reach an equation describing the curve of Item 5, a rectangular hyperbola. To correspond to that curve, v must be described in terms of $[S]$ by an equation having this general form:

$$v = \frac{a \cdot [S]}{b + [S]}$$

where a and b are constants. Unfortunately we cannot complete the solution to this form unless one or another simplifying assumption is made. One such assumption was proposed in 1913 by *Michaelis* and *Menten,* leading them to the equation that bears their names. Because their assumption has not proved to be a generally warranted one, we will delay considering it until later. The equation they reached, however, namely the ____________-____________ equation, has proved widely applicable without formal modification.

36

* It was in 1925 that a more generally valid assumption was proposed by *Briggs* and *Haldane* in a short, historic paper (see Appendix D). Briggs and Haldane pointed out an important circumstance illustrated in the accompanying figure. (For the development of this figure, see Reference 7.)

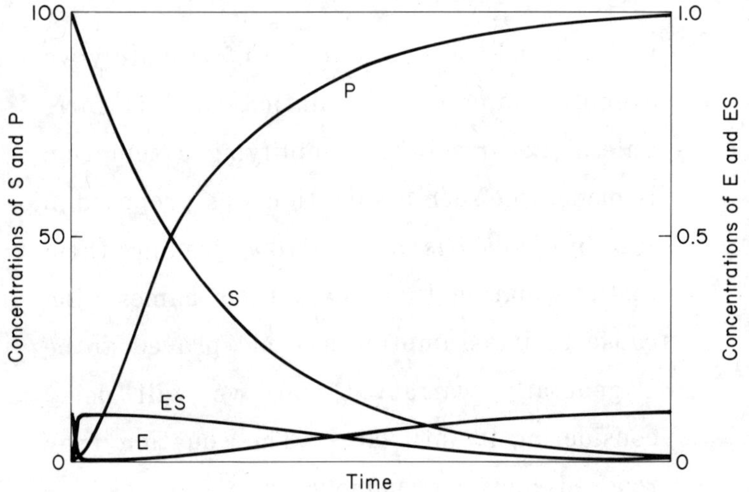

This diagram shows how the concentrations of the various components tend to behave during the course of an enzymatic reaction:

$$E + S \underset{k_{-1}}{\overset{k_1}{\rightleftharpoons}} \_\_\_\_\_\_ \longrightarrow \_\_\_\_\_\_\_\_\_\_\_\_$$

Notice that the concentrations of S and P are referred to the scale at the left, and those of E and ES to the scale at the right. Furthermore, the scale at the right has been magnified 100-fold. Accordingly, after a first, transient moment, $[S]$ and $[P]$ change very much more | rapidly | slowly |, in absolute terms, than do $[E]$ and $[ES]$.

Page 20

35

Michaelis-Menten

36

$$ES \xrightarrow{k_2} E + P$$

rapidly

37

* Indeed, as a good approximation the rate of change of $[E]$ and $[ES]$ with time may be called zero in comparison to the rate of change of either ______ or P. The assumption proposed by Briggs and ____________ is known as the *steady-state approximation.* Very soon after an enzymatic reaction begins, a steady state is reached, in which the rate of change with time of the concentration of the transient intermediate(s) is essentially zero when compared to the rate of change of the concentration of the reactants or products.

37

S

Haldane

38

* In algebraic terms, the steady-state approximation says that

$$\frac{d[ES]}{dt} = 0$$

Therefore the equation in Item 34, which was

$$\frac{d[ES]}{dt} = k_1([E_t]-[ES])[S]-k_{-1}[ES]-k_2[ES]$$

can be rewritten as follows:

$$k_1([E_t]-[ES])[S]-k_{-1}[ES]-k_2[ES] = 0$$

In order to solve this equation for $[ES]$, we will first expand:

$$k_1[E_t][S]-k_1[ES][S]-k_{-1}[ES]-k_2[ES] = 0$$

$$k_1[E_t][S] = (k_1[S]+k_{-1}+k_2)[ES]$$

Solve this equation for $[ES]$.

$$[ES] = \frac{\qquad}{\qquad}$$

39

* Divide both the numerator and denominator of the fraction by k_1.

$$[ES] = \boxed{}$$

38

$$[ES] = \frac{k_1[E_t][S]}{k_1[S] + k_{-1} + k_2}$$

40

* Recalling from Item 32 the basic rate equation for the formation of product

$$v = k_2\,[ES]$$

we can write

$$v = \boxed{}$$

39

$$\frac{[E_t]\,[S]}{[S] + \frac{k_{-1} + k_2}{k_1}}$$

41

* The expression

$$\frac{k_{-1} + k_2}{k_1}$$

appearing in the denominatcr contains only constants, and hence it can be replaced by a single constant, which we will call the *Michaelis constant* and designate K_m. On making that replacement, we have:

$$v = \boxed{}$$

40

$$\frac{k_2\,[E_t]\,[S]}{[S] + \frac{k_{-1} + k_2}{k_1}}$$

41

$$\frac{k_2 [E_t] [S]}{[S] + K_m}$$

42

* We can simplify this equation still further by asking what would happen if $[S]$ were very much greater than K_m, so that the contribution of K_m to the denominator could be neglected. The resulting equation would reduce to the simple form

$$v = \_\_\_\_\_\_\_\_$$

This equation tells us what value the velocity approaches at very high substrate concentrations.

42

$k_2 [E_t]$

43

* Such very high substrate concentrations cause the rate to approximate V_{max}, as defined in Item 5. Therefore we may write

$$V_{max} = \_\_\_\_\_\_\_\_$$

(This definition can also be reached from the basic rate equation, $v = k_2 [ES]$. When $[S]$ is infinitely high, all of the enzyme is converted to ES. Hence, $[ES]$ becomes equal to ______.)

44

* Accordingly, if we substitute V_{max} for $k_2[E_t]$ in the equation in Item 41:

$$v = \frac{k_2[E_t][S]}{[S] + K_m}$$

we obtain the Michaelis-Menten equation:

$v =$ 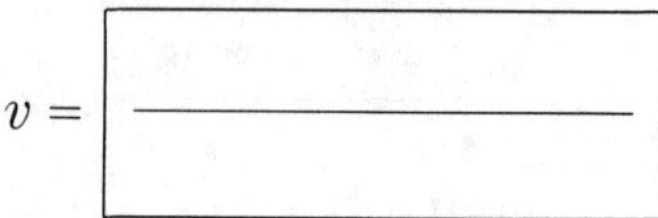

This equation is often seen an in alternative form:

$$v = \frac{V_{max}}{\frac{K_m}{[S]} + 1}$$

Either form shows that the rate is decreased from the maximal according to the magnitude of the substrate concentration in relation to the value of K_m.

43

$k_2[E_t]$

$[E_t]$

44

$$\frac{V_{max} \cdot [S]}{[S] + K_m}$$

Meaning of Constants V_{max} and K_m

45

* The quantities K_m and V_{max} in this equation are characteristic for a given enzymatic reaction under a given set of conditions. If we know these two quantities we can calculate the rate of the reaction for the full range within which the ______ ______ equation applies.

45

Michaelis-Menten

46

* Since K_m in this equation stands for the algebraic quantity, ____________, in the general case it has no simple, theoretical significance, but is merely a complex of these three rate constants. Nevertheless it and the V_{max} are the only quantities needed to describe the rate of an enzymatic reaction under a given set of conditions.

47

* Even though K_m has no simple theoretical significance, it does have significance of a practical sort. Suppose we test the rate at a substrate concentration just equal to the K_m. If $[S] = K_m$, we can write

$$v = \frac{V_{max} \cdot [S]}{[S] + [S]}$$

Simplifying this equation gives:

$$v = \frac{\qquad}{\qquad}$$

46

$$\frac{k_2 + k_{-1}}{k_1}$$

48

* This result provides us with a simple, working definition for the K_m, namely that it is numerically identical with the concentration of the substrate that will permit the catalytic reaction to go:

( ) at its maximal velocity

( ) at half its maximal velocity

47

$$v = \frac{V_{max}}{2}$$

48

at half its maximal velocity

49

* Let us review in words the meaning, in the present context, of each of the four symbols used in the Michaelis-Menten equation:

v = ______________________

$[S]$ = ______________________

K_m (The Michaelis constant) = ________

____

____

V_{max} = ______________________

____

____

50

Values for V_{max} are optimally expressed in relation to the enzyme concentration, stated in molar units, whenever the molar concentration of the enzyme is available. For example, we might say that the V_{max} for a given enzyme is 500 millimoles of substrate consumed per minute for each millimole of enzyme present. This unit of velocity can be simplified algebraically to [min^{-1} | millimoles]. That is, for the equation

$$V_{max} = k_2 [E_t]$$

if we calculate the rate hypothetically to be expected with $[E_t]$ at a unit molar quantity, V_{max} becomes identical with k_2. The term *turnover number* is used for the V_{max} expressed under that convention. To say that an enzyme has a turnover number of 10,000 per minute means that each molecule of the enzyme will catalyze the modification of 10,000 [molecules | millimoles] of substrate per minute. Turnover numbers ranging from 6 to 17,000,000 have been recorded for various enzymes.

49

v = velocity of the enzymatic reaction

$[S]$ = the substrate concentration

K_m = the substrate concentration permitting a half-maximal velocity:

$$\frac{k_2 + k_{-1}}{k_1}$$

V_{max} = the maximal velocity that can be attained by elevating $[S]$

50

min^{-1}

molecules

APPROXIMATIONS FOR HIGH AND LOW SUBSTRATE CONCENTRATIONS

51

* We can discover some qualitative features of the Michaelis-Menten equation by considering the effects predicted by it if we use extreme concentrations of the substrate.

If we use very low concentrations of substrate, so that $[S] \ll K_m$,* the equation

$$v = \frac{V_{max}[S]}{K_m + [S]}$$

can be simplified to the approximation

$$v = \frac{V_{max}[S]}{K_m}$$

The quantity $\frac{V_{max}}{K_m}$ will be a [variable|constant].

Therefore the reaction velocity will be [directly|inversely] proportional to the substrate concentration at such low concentrations. The reaction is therefore [first-order|second-order] with respect to substrate.

*To be read, "[S] is very much smaller than K_m."

52

* We may say that this linear relation prevails when $[S]$ is less than one-tenth of K_m, provided that we are satisfied with a correspondence within 10 per cent. This condition applies to part $\boxed{A \mid B \mid C}$ of the accompanying curve.

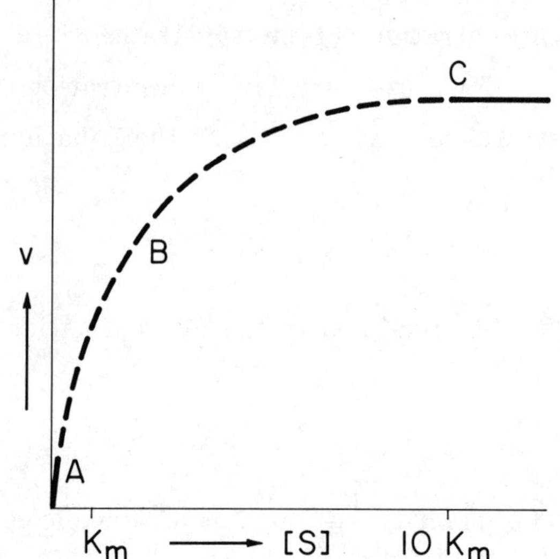

On the other hand if $[S] \gg K_m$, the Michaelis-Menten equation, ordinarily written

$$v = \boxed{\frac{\qquad}{\qquad}}$$

can be simplified to the approximation

$$v = V_{max}$$

51

constant

directly

first-order

(See Item 21 if in doubt.)

52

A

$$\frac{V_{max} \cdot [S]}{K_m + [S]}$$

(or the alternative form)

53

C

54

enzyme

53

* That is, the reaction velocity is now independent of substrate concentration, so we can say that the reaction is *zero-order* with regard to substrate concentration. We may take this condition to be approximated when $[S]$ is at least 10 times K_m. In the figure in Item 52, this condition applies in the region marked [*A* | *B* | *C*].

54

* If all cofactors, as well as the substrate, are at optimally high concentrations and if the enzyme concentration is varied through a circumscribed range, the rate may well be found experimentally to vary linearly with enzyme concentration. Thus we obtain the conditions for a(n) [enzyme | substrate] assay, as considered in Item 2.

55

* We want the changes in substrate concentration during an enzyme assay to have little or no effect on the velocity of the reaction. That result may be obtained either by observing the reaction for only a very short time interval, or more simply, by using a substrate concentration so high that the rate is nearly independent of substrate concentration. For that purpose the substrate concentration ought to be greater than the K_m by a factor of at least ____________.

56

* The Michaelis-Menten equation thus can assist us in designing enzymatic assay procedures by predicting conditions under which the velocity of a catalytic reaction will be (*a*) linear with enzyme concentration or (*b*) linear with substrate concentration. When [*S*] lies between 0.1 K_m and 10 K_m (corresponding to part | *A* | *B* | *C* | of the curve in Item 52), the foregoing simple linear approximations of rate | apply | do not hold |, and we need to use the complete equation to describe the dependence of reaction velocity on substrate concentration. This intermediate range of substrate concentrations is in fact the principal subject of this program.

55

10

56

B

do not hold

DERIVATION OF MICHAELIS-MENTEN EQUATION UNDER EQUILIBRIUM ASSUMPTION

57

* The derivation of the Michaelis-Menten equation under the steady-state approximation of Briggs and Haldane, which we completed in Item 44, provides us with the most generally applicable interpretation of the quantities included in that equation. In our preference for the derivation by Briggs and Haldane, we have delayed presenting until now the original derivation by which ________ and ________ reached the equation bearing their names.

57

Michaelis, Menten

58

* That original derivation was made under an assumption that is not so much erroneous as it is restrictive, representing a special case within the steady-state approximation. When a given catalytic process can be shown to fall within that special case, the K_m gains a particular theoretical significance. Accordingly, our reasons for now tracing the Michaelis-Menten derivation are | purely historic | to describe a special case |.

59

* If in the sequence

$$E + S \underset{k_{-1}}{\overset{k_1}{\rightleftharpoons}} ES \xrightarrow{k_2} \_\_\_\_\_\_$$

the conversion of ES to $P + E$ should happen to be much slower than its conversion back to E and S, then the reaction

$$E + S \underset{k_{-1}}{\overset{k_1}{\rightleftharpoons}} ES$$

can be considered to be at an equilibrium. That is, the forward reaction rate, $k_1 [E] [S]$, will be almost equal to the backward reaction rate, k_{-1} [______]. In algebraic terms, under this assumption

$$k_1 [E] [S] = k_{-1} [\_\_\_\_\_\_]$$

60

* Here we have the *equilibrium* assumption of Michaelis and Menten: That the reaction marked k_2 in our scheme is relatively so slow that we may consider ES to be in a state of ____________ with respect to E and S. It is worth noting that under these conditions the amount of ES present will change only slowly, and the steady-state approximation | will also | will not | be met. Hence the equilibrium assumption represents a special case within the Briggs-Haldane derivation.

58

to describe a special case

59

$P + E$

(either order)

ES
(The full equation)
$k_1 [E] [S] = k_{-1} [ES]$

60

equilibrium

will also

61

(The full equation)

$k_1([E_t] - [E_t] - [ES])[S] = k_{-1}[ES]$

62

E_t

S

(either order)

63

(The full equation)

$k_1 [E_t] [S] = (k_{-1} + k_1 [S]) [ES]$

61

* We will carry our new equation, $k_1 [E] [S] = k_{-1} [ES]$, through the same series of steps we followed in the Briggs-Haldane derivation. First we will modify the equation by noting that $[E] = [E_t] - [ES]$. By substitution we obtain

$$k_1([E_t] - [\_\_\_\_\_\_])[S] = k_{-1}[\_\_\_\_\_\_]$$

62

* Performing the indicated multiplication gives us

$$k_1[\_\_\_\_][\_\_\_\_] - k_1[ES][S] = k_{-1}[ES]$$

63

* If we gather the terms in which ES appears, we have

$$k_1[E_t][S] = (\_\_\_\_\_\_ + \_\_\_\_\_\_)[ES]$$

64

* Solving for $[ES]$ we obtain:

$$\boxed{\frac{k_1 [E_t] [S]}{\qquad}} = [ES]$$

65

* If we substitute this value for $[ES]$ into our original rate equation, $v = k_2[ES]$, we obtain:

$$v = \frac{k_1 k_2 [E_t][S]}{k_{-1} + k_1 [S]}$$

Dividing both numerator and denominator of the fraction by k_1 gives us:

$$v = \frac{\qquad}{\qquad}$$

66

* Since k_{-1}/k_1 represents the ratio between two constants, it will itself be a constant. This ratio between the rate constant for the backward reaction and the rate constant for the forward reaction will determine the position of equilibrium, and is as we have seen an *equilibrium constant.* For the reaction

$$E + S \underset{k_{-1}}{\overset{k_1}{\rightleftharpoons}} ES$$

the larger the equilibrium constant given by k_{-1}/k_1, the greater will be the degree of dissociation of ES; hence this constant is the | dissociation | formation | constant for ES.

64

$k_{-1} + k_1 [S]$

(either order)

65

$$\frac{k_2 [E_t][S]}{k_{-1}/k_1 + [S]}$$

66

dissociation

67

dissociation

$$\frac{k_2[E_t][S]}{K_m + [S]}$$

68

$$\frac{V_{max} \cdot [S]}{K_m + [S]}$$

67

* It was for this ratio, k_{-1}/k_1, that the designation K_m, the *Michaelis constant,* was originally adopted.* The term K_m was later carried over into the more general analysis of Briggs and Haldane. Therefore, wherever the equilibrium assumption has been shown to apply, K_m takes the special significance that it is the ____________ constant for the complex ES. Substituting K_m into the equation in Item 65 according to the above definition leads us to the equation

$v =$ ____________

68

* We have already seen that the product $k_2[E_t]$ has the particular significance represented by the term V_{max}. If we substitute V_{max} into our equation for $k_2[E_t]$, it takes the form

$v =$ 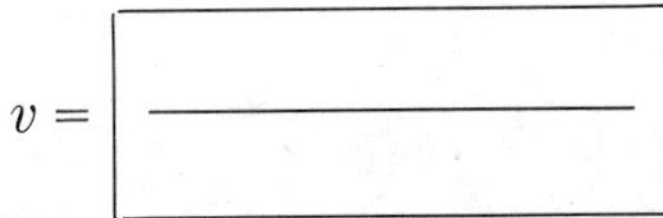

Thus the same Michaelis-Menten equation has been reached by two routes.

*Some authorities use the designation K_s when the constant in question is the dissociation constant of ES.

COMPARISON OF DERIVATIONS; MEANING OF CONSTANTS

69

* Although the two equations reached under the two different assumptions may be seen to be formally identical, the definition of K_m is different in the two cases. To emphasize the difference, complete the following table:

Derivation according to	*Essential assumption*	*Definition of K_m*
__________ (names)	That the rate of change of $[ES]$ with time is negligible	K_m = __________ (Item 41)
__________ (names)	That E and S are in equilibrium with ES	K_m = __________ (Item 66)

69

Briggs and Haldane:

$$\frac{k_{-1} + k_2}{k_1}$$

Michaelis and Menten:

$$\frac{k_{-1}}{k_1}$$

70

* Examine once more the process

$$E + S \underset{k_{-1}}{\overset{k_1}{\rightleftharpoons}} ES \xrightarrow{k_2} E + P$$

and complete the following algebraic equivalents of the two verbal statements of what is assumed for each of our two kinetic models:

Verbal statements	*Algebraic statements*
ES is formed from *E* and *S* essentially as fast as it is consumed, either by conversion back to *E* and *S* or on to *E* and *P*.	$k_1([E] \cdot [S]) = k_{-1}[\_\_\_\_] + \_\_\_\_ [ES]$
ES is formed from *E* and *S* essentially as fast as it redissociates to them.	$k_1([E] \cdot [S]) = k_{-1}[\_\_\_\_]$

The reason that the term containing k_2 does not appear in the latter equation (and hence in the corresponding definition of K_m) is of course that k_2 is taken to be | large | negligible | in magnitude, relative to k_{-1}. Hence we confirm a statement made earlier: If the equilibrium assumption holds, the steady-state approximation | will not | will necessarily | hold.

71

* In each of these two cases the derivation from the above algebraic statements proceeds by the same five steps:

1. $[E_t] - [ES]$ is introduced in place of ______

2. The resulting equation is solved for $[ES]$

3. The resultant value of $[ES]$ is introduced into the rate equation $v =$ ______ $[ES]$ (See Item 32.)

4. ______ is introduced in place of $k_2\,[E_t]$

5. _____ is introduced in place of $\frac{k_{-1}}{k_1}$ or $\frac{k_{-1} + k_2}{k_1}$

(The derivation under the steady-state approximation is shown concisely in Appendix A.)

70

(The full equations)

$k_1\,([E] \cdot [S]) = k_{-1}\,[ES] + k_2\,[ES]$

$k_1\,([E] \cdot [S]) = k_{-1}[ES]$

negligible

will necessarily

71

[E]

k_2

V_{max}

K_m

72

* Unfortunately, investigators have on many occasions assumed that because the rate of a given enzymatic reaction corresponds to the Michaelis-Menten *equation,* it must necessarily correspond to the Michaelis-Menten *assumption* (or to the Michaelis-Menten *model*). That is to say, they assume that the K_m obtained represents the dissociation constant of the ____________________, and hence that it measures the affinity of E for S. There is no compelling reason for believing *a priori* that this assumption will be correct for a given case; it may be awkward to verify; and indeed it is not true for a number of carefully studied cases of enzyme action.

73

* For example, for horseradish peroxidase it has been possible to measure $[ES]$ photometrically in the system

$$E + S \underset{k_{-1}}{\overset{k_1}{\rightleftharpoons}} ES \xrightarrow{k_2} E + P$$

This measurement has permitted the demonstration that in this case k_2 is 26 times as large as k_{-1}. Notice that if $k_2 \gg k_{-1}$, then

$$\frac{k_{-1} + k_2}{k_1} \gg \frac{k_{-1}}{k_1}$$

Is it possible to assume that the K_m of this enzymatic reaction is the dissociation constant of ES? ______ Should one in general take the Michaelis constant to represent the dissociation constant of the enzyme-substrate complex for any given catalytic process where no evidence is available to the contrary?

______

72

enzyme-substrate complex

73

No. (Calculation shows that the K_m in this case will be $27 \times k_{-1}/k_1$.)

No

74

Michaelis, Menten
half-maximal

74

* To avoid the usually unjustified acceptance of the equilibrium assumption of ____________ and ____________, it will be worth our while to emphasize the generalized definition of the Michaelis constant, that it represents merely the concentration of the substrate able to produce a ____________________ rate in the enzymatic reaction. Through the remainder of this program we will be treating the general case; and hence we can accord K_m only this limited significance.

GRAPHIC DETERMINATION OF K_m AND V_{max}

75

* As with the Henderson-Hasselbalch equation in the program on *pH and Dissociation,* our ready comprehension of the simple kinetic equation of Michaelis and Menten:

$v =$ 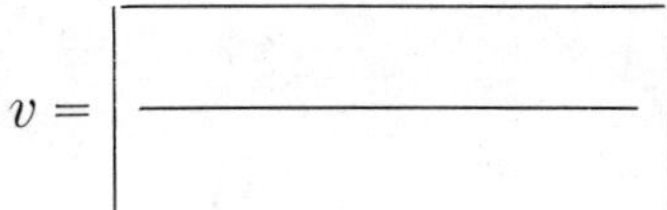

depends on our securing practice in its use.

76

* Suppose that glucose can be converted to another substance by a certain enzyme. The table below shows for this hypothetical case the rates produced at several glucose concentrations:

Sugar concentration (M)	*mmoles per l. converted per min for each mmole of enzyme present per l.*
1×10^{-5}	150
2×10^{-5}	256
1×10^{-4}	600
3×10^{-4}	770
5×10^{-4}	818

Note that the units in which the rate is expressed can be simplified algebraically to mmoles liter^{-1} min^{-1}. The value we obtain for V_{max}, expressed in the same terms, can be called

( ) a dimensionless number

( ) a turnover number

75

$$\frac{V_{max} \cdot [S]}{K_m + [S]}$$

(a variant form may be given:

$$v = \frac{V_{max}}{1 + \frac{K_m}{[S]}}$$
)

76

min^{-1}

a turnover number

77

* Introduce these rates on the graph. Join the points with a smooth curve.

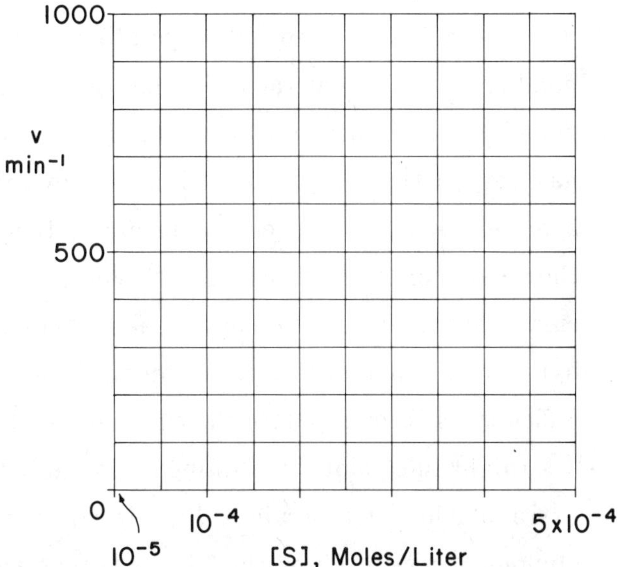

Does the shape of the curve suggest that the reaction is tending to reach a maximal rate, and that eventually concentrations will be reached above which the rate cannot be further increased? ______ By inspection of this curve one can determine | only approximately | precisely | what this maximal rate will be.

78

* If this curve is a rectangular hyperbola (as would be required for equations having the general form of the Michaelis-Menten equation) it will reach a horizontal asymptote corresponding to V_{max} at a precisely determined velocity. The difficulty is that the human eye has very little ability to estimate how high a plateau will be reached by the curve. Indeed, the eye cannot even recognize whether the curve has the true form of a rectangular hyperbola. What one does to deal with this difficulty is to manipulate the equation so that a straight-line plot is obtained instead. The extrapolation of a straight line is of course obvious. Check each of the "linear transformations" of the Michaelis-Menten equation below to see if it does indeed correspond to the original equation.

$$y = a + (b \cdot x)$$

1. $\frac{1}{v} = \frac{1}{V_{max}} + \left(\frac{K_m}{V_{max}} \cdot \frac{1}{[S]}\right)$ Yes | No

2. $\frac{[S]}{v} = \frac{K_m}{V_{max}} + \left(\frac{1}{V_{max}} \cdot [S]\right)$ Yes | No

3. $v = V_{max} - \left(K_m \cdot \frac{v}{[S]}\right)$ Yes | No

Clue: Begin with the form

$$v = \frac{V_{max}}{\frac{K_m}{[S]} + 1}$$

and solve for V_{max}. (In case of difficulty, see Appendix B.)

Page 46

77

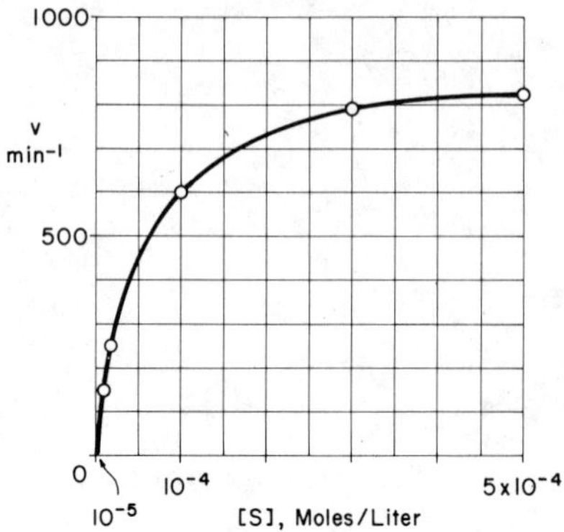

Yes

only approximately

78

Yes

Yes

Yes

79

Yes

$[S]$

$\frac{v}{[S]}$

79

* Since the equation of a straight line is $y = a + bx$, look at the first equation and decide whether a plot of $1/v$ against $1/[S]$ should give a straight line. | Yes | No | Referring to the second equation, against what variable should one plot $[S]/v$ to obtain a straight line? ______ Against what quantity should one plot v in the third equation to obtain a straight line? ______

80

* All three of these methods are used for plotting kinetic results. The most widely known plot (although probably not the most satisfactory*) is the one using the first of these equations, and is known as the *Lineweaver-Burk* plot. Since the reciprocal of the velocity is plotted against the reciprocal of the substrate concentration, this plot is often referred to as a *double-reciprocal* plot. Complete the table below (from Item 70) to obtain the quantities to be plotted. (Reciprocals are easily obtained with a slide rule.)

$[S]$ M	$1/[S]$ M^{-1}	v min^{-1}	$1/v$, min
1×10^{-5}	______	150	______
2×10^{-5}	______	256	______
1×10^{-4}	______	600	______
3×10^{-4}	______	770	______
5×10^{-4}	______	818	______

*For a recent demonstration of the defect, see Reference 8.

81

* Plot these results on the following grid, and use a straight-edge to draw the best line to represent the points.

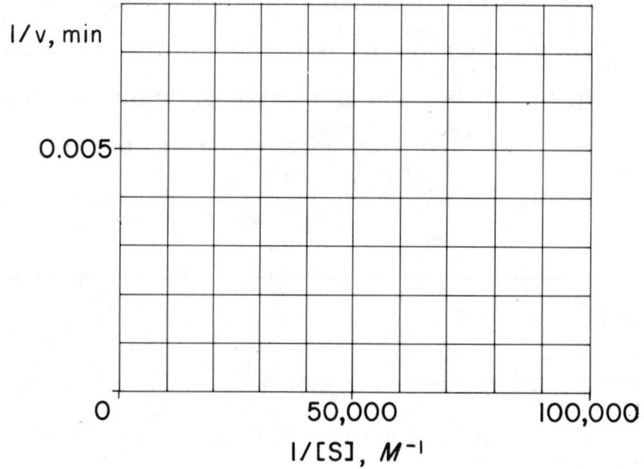

82

Compare the degrees of separation of the points representing the two highest substrate concentrations on the original plot and on the double-reciprocal plot. Which would therefore have the greater influence on the position of the line obtained, an aberration in the rate observed at 1×10^{-5} *M*, or an aberration in the rate observed at 3×10^{-4} *M*? At __________ *M*. Note further that the rates observed at lower concentrations are inherently subject to the larger percentage errors. This feature of the Lineweaver-Burk plot is leading many investigators to prefer the method of Augustinsson or that of Woolf for plotting kinetic data. (See Appendix B.)

80

$1/[S]$
1×10^5 (100,000)
5×10^4 (50,000)
1×10^4 (10,000)
3.3×10^3 (3,300)
2×10^3 (2,000)

$1/v$
0.00667
0.00391
0.00167
0.00130
0.00122

81

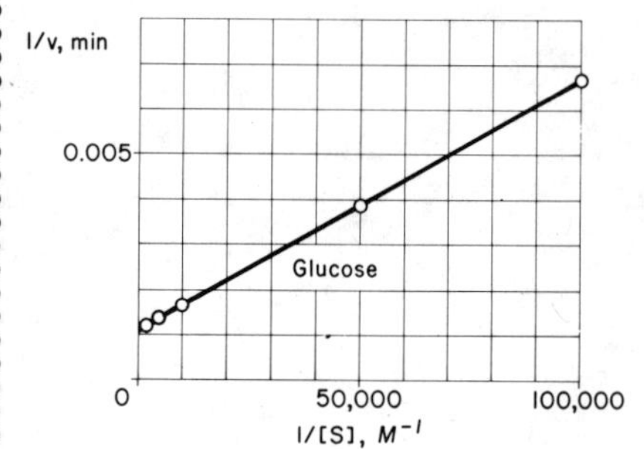

82

$1 \times 10^{-5}\ M$

83

* Since

$$v = \frac{V_{max} \cdot [S]}{K_m + [S]}$$

and

$$\frac{1}{v} = \frac{1}{V_{max}} + \left(\frac{K_m}{V_{max}} \cdot \frac{1}{[S]}\right)$$

are equivalent equations, the fact that the five points of Item 81 correspond to a straight line (thus satisfying the second equation) proves that velocity increases with substrate concentration in the manner required by the Michaelis-Menten equation. We thus accomplish the first purpose of the Lineweaver-Burk plot, namely to check by the linearity of the plot how well the reaction in question corresponds to that equation. This correspondence indicates that the substrate forms a complex with the enzyme, which throughout the period of each observation remains:

( ) at a relatively constant concentration

( ) in a state of equilibrium with respect to free glucose and free enzyme

84

* The second purpose of the Lineweaver-Burk plot, once we are satisfied with its linearity, is to evaluate V_{max} and K_m. The intercept on the y axis of the straight line of Item 81 corresponds to the value a in the equation $y = a + bx$. For our equation

$$\frac{1}{v} = \frac{1}{V_{max}} + \left(\frac{K_m}{V_{max}} \cdot \frac{1}{[S]}\right)$$

this intercept has the algebraic value ______.

83

at a relatively constant concentration

85

* Note that a zero value for $1/[S]$ corresponds to an infinitely high concentration of S, the condition at which the velocity is maximal. From the value of the intercept in the plot in Item 81, calculate V_{max} for the reaction: ______ min^{-1}.

84

$1/V_{max}$

86

* The same equation

$$\frac{1}{v} = \frac{1}{V_{max}} + \left(\frac{K_m}{V_{max}} \cdot \frac{1}{[S]}\right)$$

tells us that the slope of the line in the above figure will be K_m/V_{max}, since this value occupies the place of b in the general equation, $y = a + bx$. Hence $K_m = V_{max} \cdot \mathit{slope}$. Since the line rises by $(0.0067 - 0.0011)$ or 0.0056, as $1/[S]$ increases by 100,000, the slope is 0.0056/100,000 or 5.6×10^{-8}. Therefore $K_m =$ ____________ M.

85

900 (or thereabouts)

86

5×10^{-5}

$(5.6 \times 10^{-8} \times 900)$

87

This value is actually available graphically in the plot you prepared in Item 81. If we use a straight-edge to extend the straight line to negative substrate concentrations, it will intersect the abscissa at $-1/K_m$. Applying this test to the plot of Item 81 (below) gives a value for $-1/[S]$ of ____________, corresponding to a K_m of ____________ M. Check the result for similarity to that obtained by calculation in Item 86. We can see why this method works, as follows: When $1/v$ becomes zero, the equation

$$\frac{1}{v} = \frac{1}{V_{max}} + \left(\frac{K_m}{V_{max}} \cdot \frac{1}{[S]}\right)$$

becomes

$$-\frac{K_m}{V_{max}} \cdot \frac{1}{[S]} = \frac{1}{V_{max}}$$

Multiplying through by V_{max}

$$-K_m \cdot \frac{1}{[S]} = 1$$

$$-1/[S] = \_\_\_\_\_\_$$

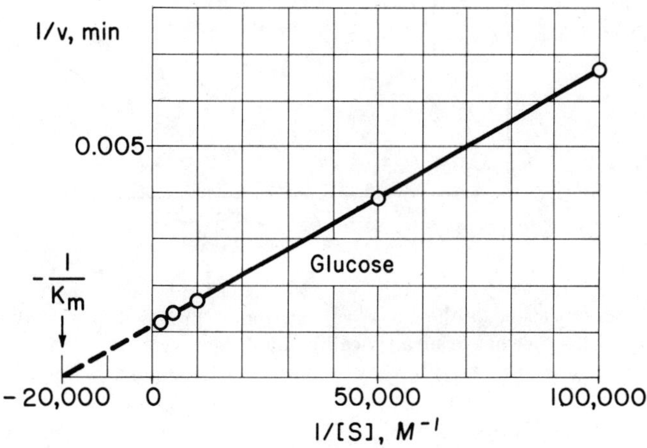

88

The disadvantage noted in Item 82 that the Lineweaver-Burk plot distributes weight inappropriately among the several observations, is leading investigators to an increasing preference for either the Hofstee* plot of v against $v/[S]$ or the Woolf plot of $[S]/v$ against $[S]$. Of these two, the Hofstee plot is intuitively more comprehensible. The quantity $v/[S]$ is the *relative rate* of the reaction, i.e., the rate for unit concentration.† If the reaction rate increases directly with substrate concentration (i.e., if substrate fails to saturate the reaction), the relative rate is constant, whatever the absolute rate. Illustrate that condition by drawing an appropriate line on this graph.

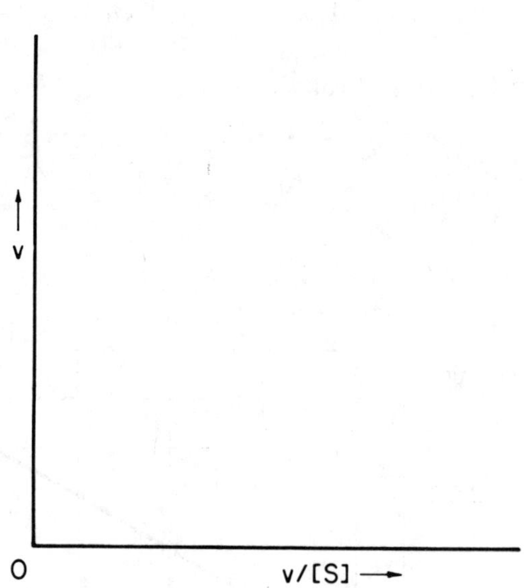

*This plot is associated with the names of both Augustinsson and Hofstee, although Woolf originally proposed it also. We will somewhat arbitrarily call it the Hofstee plot.

†The term *relative velocity* may also be applied to the rate observed, relative to the maximal velocity.

$$\sigma = \frac{v}{V_{max}}$$

87

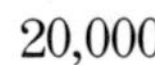

20,000

5×10^{-5}

$-1/K_m$

88

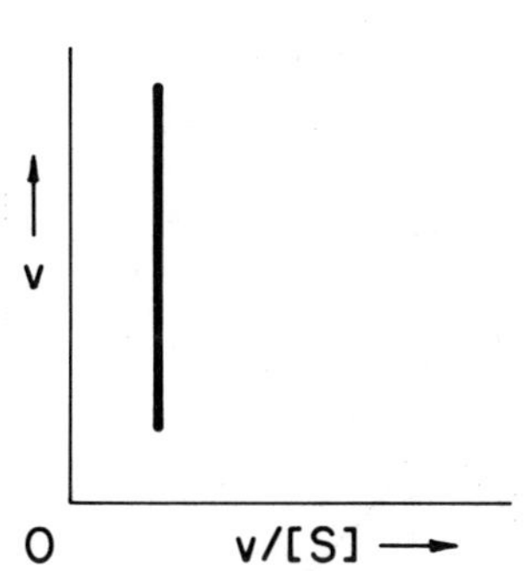

(The straight, vertical line shows that $v/[S]$ is independent of v.)

89

For a catalytic reaction (or any other reaction whose rate comes to be limited by the concentration of a second component), as we raise the substrate concentration, even though we increase the *absolute velocity*, we *decrease* the *relative velocity*. In Item 79 you showed that a plot of v versus $v/[S]$ will be a straight line, as illustrated here. Finally, as an infinite substrate concentration is reached, the relative rate becomes zero. At this point the absolute velocity will be maximal. Hence V_{max} is given by the intercept of our line on the [abscissa / ordinate]. Mark that point on the graph.

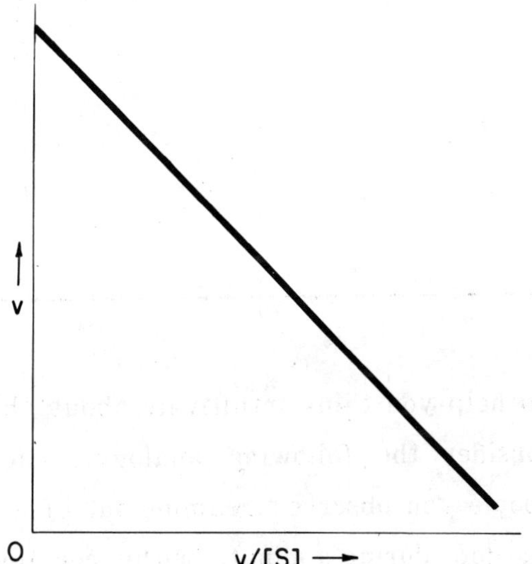

90

If we divide the value of the intercept on the ordinate by the value of the intercept on the abscissa, we get the slope of the line. This slope is negative. From the equation,

$$v = V_{max} - \left(K_m \cdot \frac{v}{[S]}\right)$$

we see that on changing its sign this ratio gives us the value of ______.

91

To help you think intuitively about this plot consider the following analogy: The more people you observe streaming out of a revolving door during a minute, within certain limits, the | larger | smaller | the proportion of those who wanted to get through in that minute actually succeeded. Or, conversely, the longer the delay of the crowd waiting to get through, the larger the number presumably emerging per minute.

89

ordinate

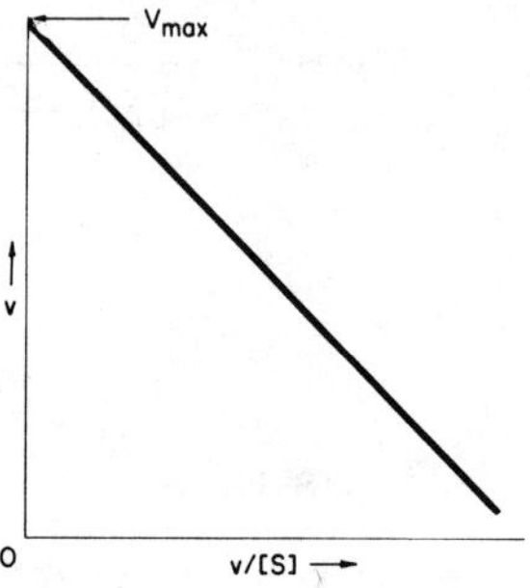

90

K_m

91

smaller

92

On the accompanying grid, plot the rate observations of Item 76, which are reproduced here:

[*Glucose*] *M*	*v* *min*$^{-1}$	*v*/[*S*] *min*$^{-1}$ *M*$^{-1}$
1×10^{-5}	150	150×10^5
2×10^{-5}	256	128×10^5
1×10^{-4}	600	60×10^5
3×10^{-4}	770	26×10^5
5×10^{-4}	818	16×10^5

Determine V_{max} and K_m from the plot.

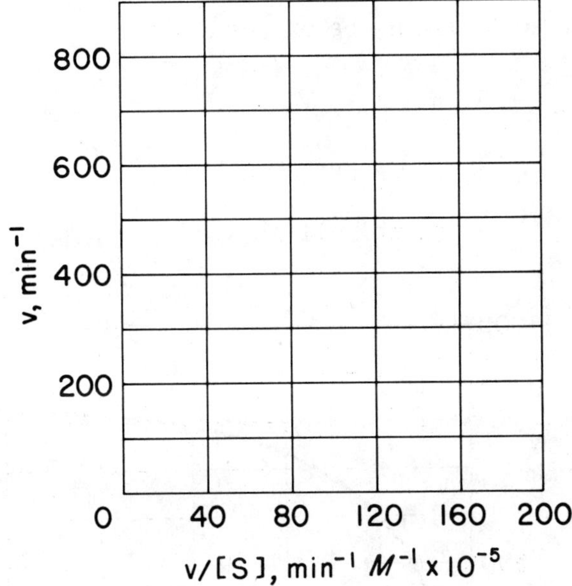

V_{max} = ______________________

K_m = ______________________

93

* Shown again below is the direct plot of the rate of the enzymatic reaction as a function of the substrate concentration, from Item 77 Draw across it a horizontal line at a level corresponding to the rate you have obtained for the maximal rate. Does this level look reasonable for the maximum suggested by the rectangular hyperbola? ______ Draw a vertical line to represent the K_m value you have just determined. Does it seem to represent a concentration of glucose that might reasonably produce a half-maximal reaction rate? ______ From that conclusion decide whether in this case K_m should be expressed in:

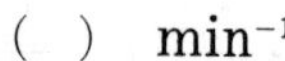

( ) min^{-1}

( ) moles per liter

( ) no units at all

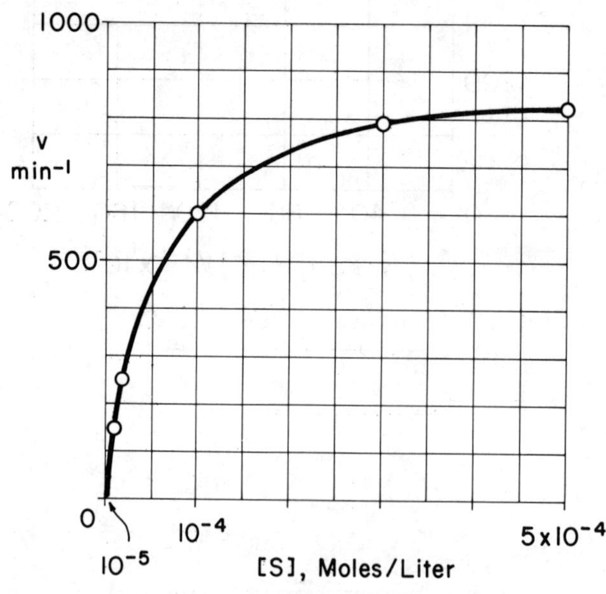

92

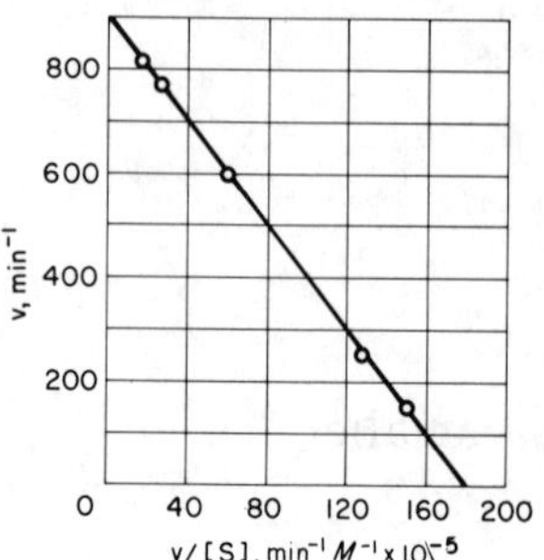

$V_{max} = 900\ min^{-1}$

$$K_m = \frac{900}{180 \times 10^5}$$

$$= 5 \times 10^{-5}$$

93

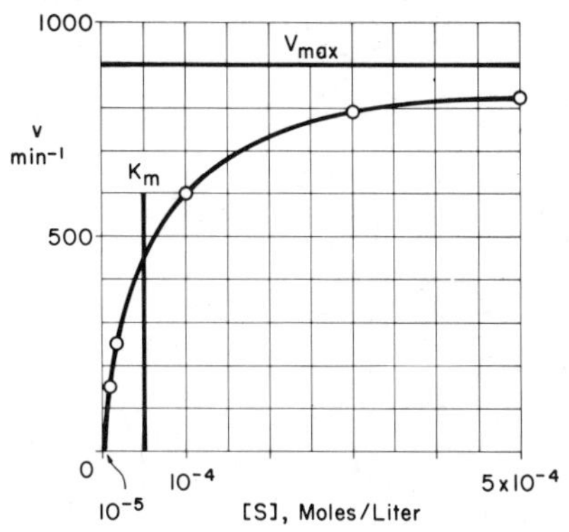

Yes

Yes

(Admittedly one's ability to estimate V_{max} is poor; otherwise one would not trouble with the linear transformation.)

moles per liter

(Recall that the K_m represents the *concentration* of S producing a half-maximal rate.)

94

If our purpose in plotting rate observations as we just did is to determine whether or not an enzymatic reaction follows the Michaelis-Menten equation, we will want our observations to describe the curvilinear part of the expected rectangular hyperbola. Which of the following distributions of four values of $[S]$ would serve this purpose better, *a* or *b*?

(*a*) Two below $K_m/10$; two above 10 K_m

(*b*) All four between $K_m/10$ and 10 K_m; two below and two above K_m.

(Refer to the graph from Item 52, repeated below, if you have difficulty in visualizing these regions.)

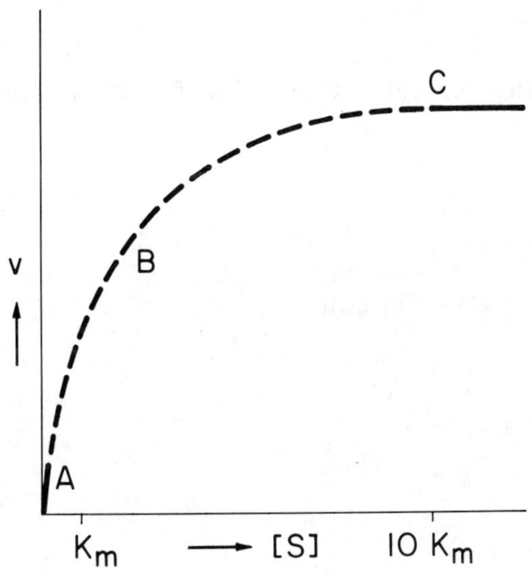

Were the values of $[S]$ for glucose used above suitable for the purpose? ______

95

As a small check on our understanding of the meaning of K_m and V_{max}, let us suppose that an enzyme reaction known to follow the Michaelis-Menten equation shows the following velocities under the conditions tabulated:

[S]	Velocity, min^{-1}	
M	at 30°	at 37°
1×10^{-5}	253	–
3×10^{-5}	378	–
1×10^{-3}	500	850

By inspection one can estimate a value for V_{max} of:

( ) slightly above 500 min^{-1}

( ) slightly above 850 min^{-1}

( ) much higher than either of these values

The corresponding value for K_m is about:

( ) 1×10^{-5} M

( ) 3×10^{-5} M

( ) 255 min^{-1}

94

b

Yes. (The highest one was at 10 K_m.)

95

slightly above 500 min^{-1} (at 30°) (The first two rates at 30° can be seen to lie in the central (*B*) portion of the Michaelis-Menten curve. Hence the much higher, third concentration must lie in region *C*, where the velocity approaches V_{max}. The data for 37° are insufficient for an estimate of V_{max} at that temperature.
The mere fact that the rate is higher at 37° does not make 850 min^{-1} a better value.)

1×10^{-5} *M* (This concentration is able to produce about half that maximal velocity; hence it must be close to K_m in value.)

96

Use the Michaelis-Menten equation to predict the velocity at 3×10^{-5} *M*. ____________. Does the result tend to confirm our estimates? ____________

The figure below shows how our three rate observations (if they are accurate) must fall on the Michaelis-Menten curve.

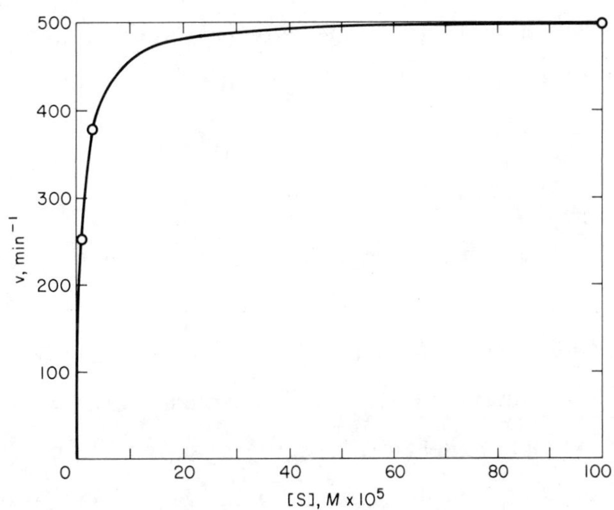

96

$$\frac{500 \times 3}{1 + 3} = 375 \text{ min}^{-1}$$

Yes

TWO-SUBSTRATE KINETICS

97

* In this development we have ignored one complication. Most real enzymes catalyze reactions into which more than one substrate enters, and by which more than one product is formed. For example,

$$A + B \rightleftharpoons P + Q$$

In these cases we might well expect the kinetic equations to be more complicated than those considered here.

Nevertheless, if the concentration of B, for example, is held constant, the velocity often proves to be a hyperbolic function of the concentration of A; i.e., the Michaelis-Menten equation still describes the relation.

If the observations are repeated at a different concentration of B, the rates will be different

but, nevertheless, an identical	and a different

hyperbola will be obtained. It is therefore apparent that under these conditions the kinetic parameters V_{max} and K_m observed for A

are, as usual,
are not

fundamental constants.

97

and a different

are not

98

infinite

98

* Only when the concentration of B is raised very high so as not to be limiting will we obtain values for the K_m and V_{max} of A that are independent of $[B]$. Therefore, to obtain these true limiting values of K_m and V_{max} for A we should need to extrapolate graphically to | infinite | zero | concentrations of B, and *vice versa*. The needed analysis appears, however, to fall beyond the scope of this primer. Nevertheless, it is important to note that the Michaelis-Menten equation tends still to apply.

(The kinetics of two-substrate reactions have proved especially valuable for discovering the mechanism of the reaction; for example, must one substrate react with the enzyme before the other, or may the order be random?)

An Example of Non-Correspondence

99

Here is another set of experimental measurements of the velocity of a different enzymatic reaction, plotted against substrate concentration. On inspection these results might appear to correspond to the Michaelis-Menten equation. To check that impression, extract the plotted values from the graph and replot them by the Lineweaver-Burk method on the blank graph. (See Item 81 if you are in doubt about the method of plotting.)

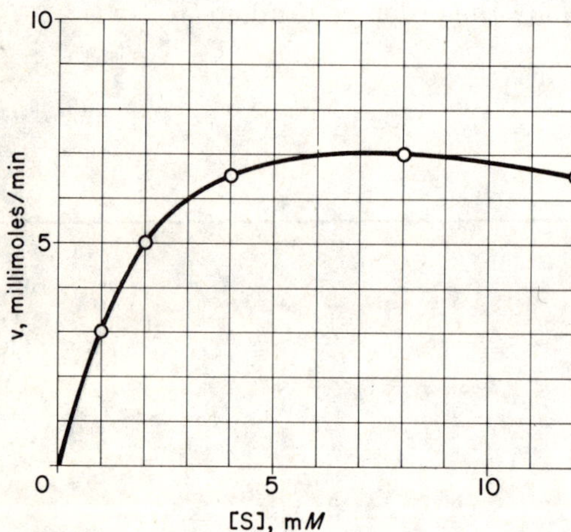

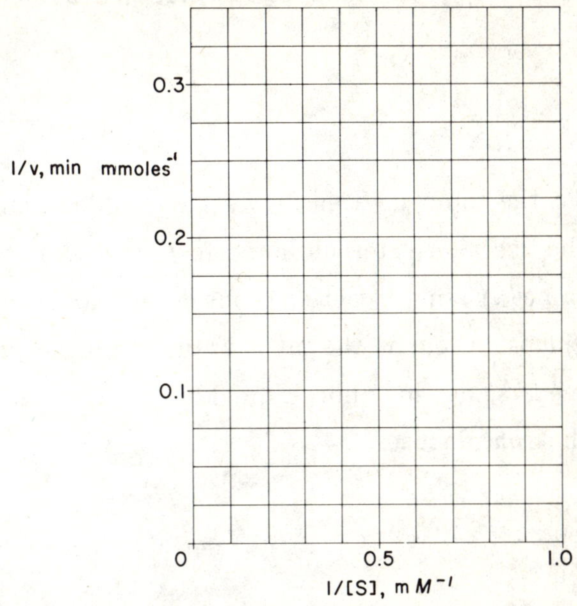

Conclusion: The rate of this reaction [does] [does not] correspond well to the Michaelis-Menten equation. The reaction probably:

( ) is represented adequately by

( ) is more complex than

the formulation $E + S \rightleftharpoons ES \longrightarrow E + P$. To explain the form of this curve, would it be plausible to suggest that an excess of substrate has inhibited the reaction in some unexpected way? ______

K_m and V_{max} for Analogous Substrates

99

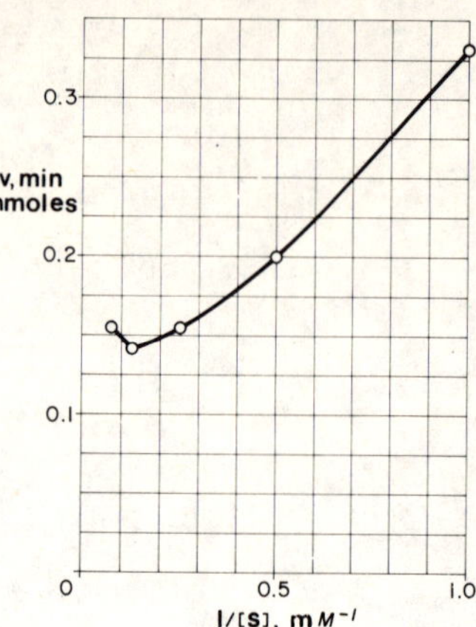

does not

is more complex than

Yes

100

* To the table in Item 80 we have added below the reaction rates obtained for mannose in the same enzyme reaction considered above for glucose. Clearly the attack on mannose proceeds somewhat more [quickly | slowly] than that on glucose.

Sugar concentration M	$\frac{1}{[S]}$	v for glucose min^{-1}	v for mannose min^{-1}	$\frac{1}{v}$ for mannose min
1×10^{-5}	100,000	150	82	0.0122
2×10^{-5}	50,000	256	150	0.0067
1×10^{-4}	10,000	600	450	0.0022
3×10^{-4}	3,300	770	670	0.0015
5×10^{-4}	2,000	818	750	0.0013

100

slowly

101

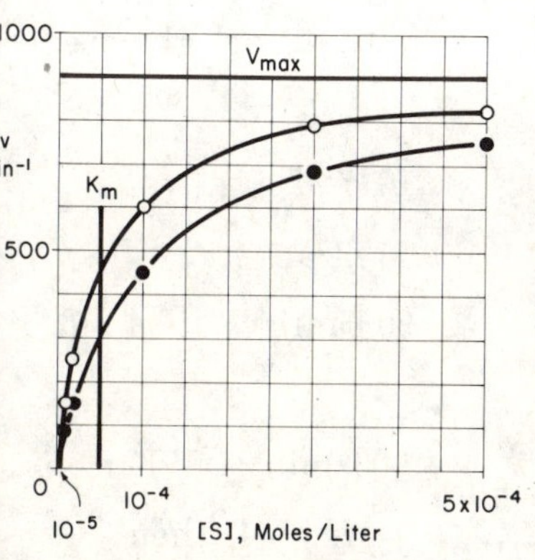

rather similar to that of glucose

101

* In the direct plot of velocity versus glucose concentration below, add the points to represent mannose. The completed plot suggests that mannose has a V_{max} that is:

( ) very much different from that of glucose

( ) rather similar to that of glucose

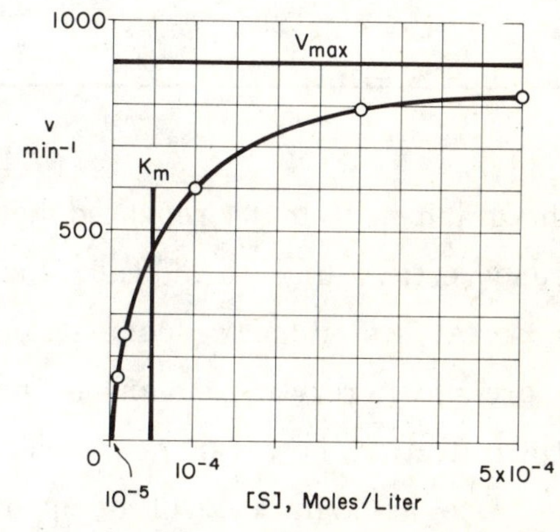

102

* Plot below $1/v$ for mannose against the reciprocal of the mannose concentration.

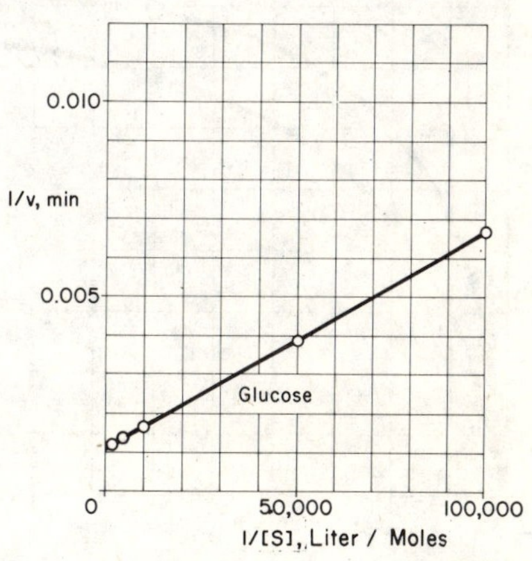

103

* Use the plot to determine the values for V_{max} and K_m for mannose in this enzymatic reaction.

V_{max} = ____________

K_m = ____________

The turnover number is ____________.

The value of k_2 is ____________.

104

* On the graph in Item 101, repeated below, add a second vertical line to indicate the K_m for mannose. At first it may seem strange that the more slowly reacting sugar has the *higher* K_m. On reflection, however, one sees that if it takes twice as high a level of mannose to produce the same half-maximal rate, it must be the [more | less] reactive substrate.

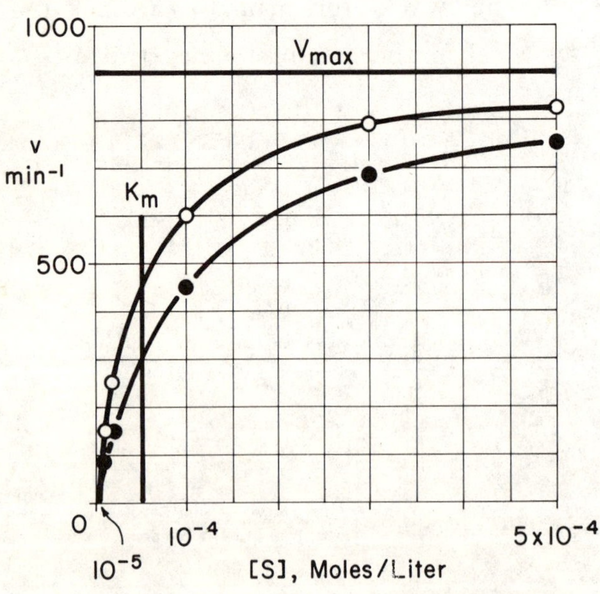

102

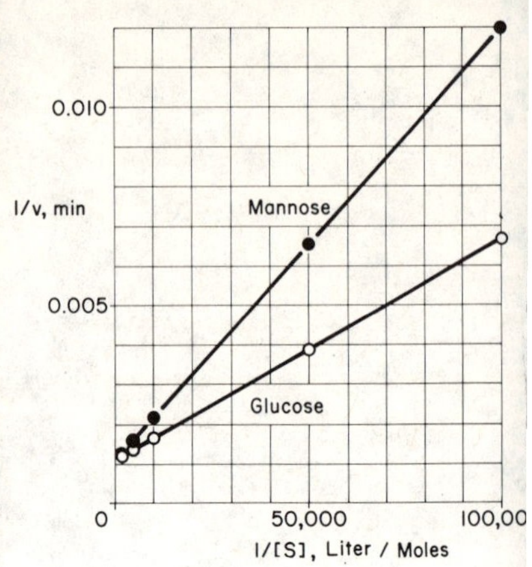

103

900 min^{-1} (The intercept is practically identical with that for glucose.)

10^{-4} *M*

900 min^{-1} (*Item 50*)

900 min^{-1} (When $[E_t]=1$, then $V_{max}=k_2$.)

104

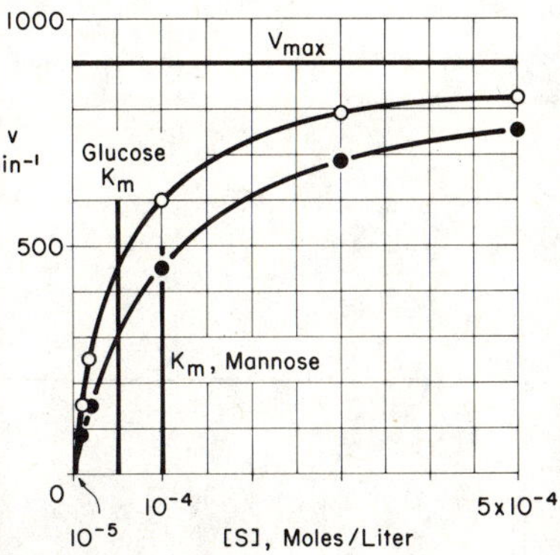

less

105

larger

smaller

B

No

105

* This feature of the K_m needs to be kept in mind: Other factors being unchanged, the higher the K_m, the [larger | smaller] the concentration needed to produce a half-maximal rate, and the [greater | smaller] is the reactivity of the substrate. Suppose that the concentration of ornithine in a given locus in a hepatic cell tends to be 10^{-4} *M*. Suppose there is also present an enzyme *A*, for which ornithine has a K_m of 10^{-3} *M*, and a second enzyme *B*, for which it has a K_m of 10^{-5} *M*. Under these conditions, we should expect ornithine to undergo catalytic modification mainly by enzyme [*A* | *B*], assuming that the two enzymes have somewhat similar values for V_{max}.

Will an event that increases the concentration of ornithine ten-fold appreciably accelerate its attack by enzyme *B*? ______

BEHAVIOR OF COMPETING SUBSTRATES

106

* Suppose we had studied the rate of the reaction under discussion, with *both glucose and mannose and the single enzyme in the same solution.* Keeping in mind the rate equation, $v = k_2[ES]$, could we expect each sugar to be modified just as quickly as if it were the only one present? ______

107

* We now have a mixture of two enzyme substrate complexes:

$$E + M \rightleftharpoons EM \longrightarrow E + P_1$$

$$E + G \rightleftharpoons EG \longrightarrow E + P_2$$

Accordingly the presence of each sugar will slow the enzymatic attack on the other. We will call mannose a *competitive inhibitor* of the modification of glucose, and glucose a *competitive inhibitor* of the modification of mannose. The term *competitive inhibitor* has the specific meaning that the inhibitory action can eventually be fully relieved by raising sufficiently the substrate concentration. The inhibitory action is relieved because at an infinite concentration of one sugar, e.g., glucose, all of the enzyme will be in the form of the enzyme-glucose complex, so that the parasitic influence of the other sugar in trapping some of the enzyme molecules will be [maximal | overcome].

106

No

107

overcome

108

* Could a substance that combined reversibly with the enzyme at the same reactive site as the substrate

$$E + I \rightleftharpoons EI$$

serve as a competitive inhibitor even if it were incapable of being modified to a product by the enzyme? ______

108

Yes

109

* Indeed most of the enzyme inhibitors having the greatest value for artificially controlling biological processes have this characteristic of combining reversibly at the reactive site of the enzyme without being capable of further modification. Therefore, that situation may be regarded as typical. Such inhibitors have the advantage that they are not destroyed by the enzyme; hence their action may be quite persistent. A classic case is the enzymatic dehydrogenation of succinate to form fumarate:

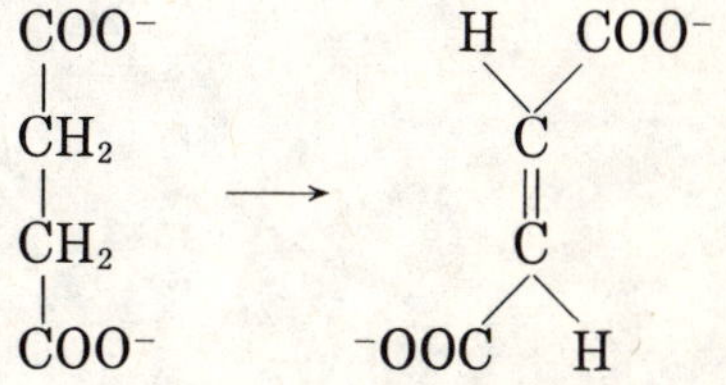

$$\begin{array}{c} COO^- \\ | \\ CH_2 \\ | \\ CH_2 \\ | \\ COO^- \end{array} \longrightarrow \begin{array}{c} H \quad COO^- \\ \diagdown \diagup \\ C \\ \| \\ C \\ \diagup \diagdown \\ ^-OOC \quad H \end{array}$$

Malonate, $CH_2(COO^-)_2$, is a potent competitive inhibitor of this enzymatic reaction. Can malonate be dehydrogenated in the way succinate is dehydrogenated? ______

110

A constant, K_i, somewhat analogous to K_m, has been introduced to describe the effectiveness of a substance in inhibiting a given enzymatic reaction. The larger the value of K_i, the higher the concentration of the inhibitor required to produce a given degree of slowing of the reaction.

Could one explore in detail the chemical nature of the binding structures at the active site of the enzyme by comparing the K_i values of a series of such inhibitory analogs, varied systematically in their structures? ______ Such an exploration has been called an *inhibition analysis.*

109

No

110

Yes

DIFFERENTIATION BETWEEN COMPETITIVE AND NON-COMPETITIVE INHIBITION

111

* The method we use for determining whether an inhibitor is *competitive* or *non-competitive* is to determine the reaction rate at various substrate concentrations in the presence of a constant level of the inhibitor. Because $[ES]$ will still tend to remain constant during the period of observation, the Lineweaver-Burk plot will still yield a straight line. Since the effect of the competitive inhibitor is abolished when the concentration of the substrate becomes infinite, the V_{max} will be | the same as | much smaller than | it is in the absence of the inhibitor. This means that the intercept on the ordinate | will | will not | be changed by the presence of the inhibitor.

111

the same as

will not

112

* Because the influence of competitive inhibitors is abolished at high substrate concentration, the V_{max} is independent of the presence of such inhibitors. A larger amount of substrate is needed, however, to convert the enzyme to the enzyme-substrate complex when the inhibitor is present, and thus the apparent K_m is:

( ) increased

( ) decreased

( ) also unchanged

113

* On the other hand, if the inhibitor combines with the enzyme in such a way that it does not tend to be displaced even by infinitely high concentrations of the substrate, i.e., if it acts as a ________________ inhibitor, the V_{max} will be:

( ) increased
( ) decreased
( ) unchanged

by the presence of the inhibitor.

112

increased

114

* Non-competitive inhibitors can be thought of as removing a certain percentage of the enzyme molecules permanently from the reaction — the remaining enzyme molecules behaving normally. This decrease in the total amount of active enzyme present is what leads to the | increase | decrease | in the V_{max}. Since the remaining portion of the enzyme behaves normally, and since the value of K_m does not depend on enzyme concentration, the K_m | is decreased | remains unchanged |.

113

non-competitive

decreased

114

decrease

remains unchanged

115

* In light of these conclusions, complete the following table by inserting the words *increased, decreased,* or *unchanged* as needed:

Type of inhibition	*Effect on V_{max}*	*Effect on the apparent K_m*
Competitive	__________	__________
Non-competitive	__________	__________

116

* The following table shows the rates at which a given substrate enters an enzymatic reaction (*a*) in the absence of any inhibitor, and (*b*) and (*c*) in the presence of a constant amount, respectively, of each of two inhibitors. First plot the data directly, *v* against [*S*], on the grid below. Draw smooth curves to join the points for each case. By examination of the table of the preceding item, case (*b*) appears by inspection perhaps to represent [competitive | noncompetitive] inhibition, and case (*c*), more likely to represent ____________ inhibition.

[*S*] m*M*	*Velocity* (*a*) *sec*$^{-1}$	*Velocity* (*b*) *sec*$^{-1}$	*Velocity* (*c*) *sec*$^{-1}$
1	2.5	1.17	0.77
2	4.0	2.10	1.25
5	6.3	4.00	2.00
10	7.6	5.7	2.50
20	9.0	7.2	2.86

v sec^{-1} (0, 2, 4, 6, 8, 10)

[S], m*M* (0, 10, 20)

(The difficulty you will encounter in these decisions serves to demonstrate the advantage of the linear plot.)

115

Effect on V_{max}

unchanged

decreased

Effect on K_m

increased

unchanged

116

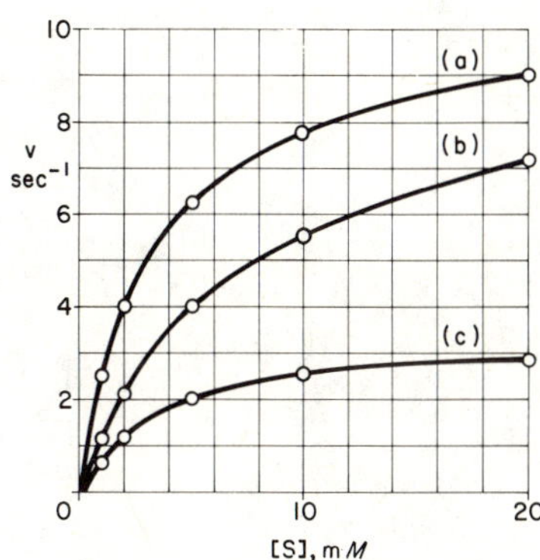

competitive (since K_m is changed, and V_{max} appears perhaps not to be)

non-competitive (since V_{max} is changed, and K_m appears perhaps not to be)

117

* Apply to these cases the objective test for competitive or non-competitive inhibition by plotting the data on the grid below according to the method of Lineweaver and Burk. Extend the ordinate or abscissa scale as needed. Applying the criteria in Item 115 label the lines for (*b*) and (*c*) either *competitive* or *non-competitive.*

$\frac{1}{[S]}$	(*a*) $1/v$	(*b*) $1/v$	(*c*) $1/v$
1.0	0.40	0.86	1.30
0.5	0.25	0.48	0.80
0.2	0.16	0.25	0.50
0.1	0.13	0.18	0.40
0.05	0.11	0.14	0.35

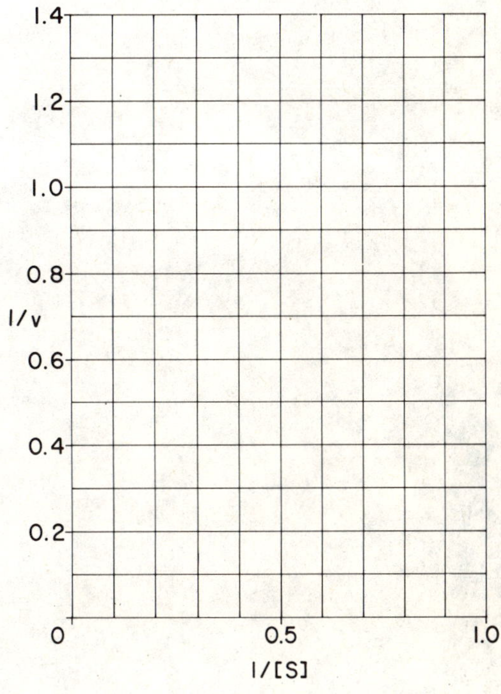

118

***** The plot just completed shows that the V_{max} of the uninhibited reaction is ______ millimoles sec^{-1}, and the K_m, ______ mM. We can see by looking at the upper line on the graph that in case (*c*) the action of the inhibitor:

( ) disappears when a very large excess of substrate is added

( ) does not disappear even when a very large excess of substrate is added

117

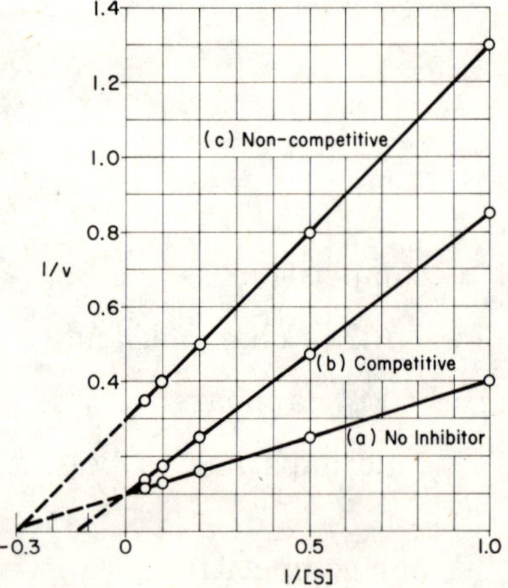

118

10

3

does not disappear even when a very large excess of substrate is added

119

Replot the observations of Item 116 (presented again below) by the Augustinsson method on the accompanying grid.

[S] mM	v_a sec^{-1}	$v_a/[S]$	v_b sec^{-1}	$v_b/[S]$	v_c sec^{-1}	$v_c/[S]$
1	2.5	_____	1.17	_____	0.77	_____
2	4.0	_____	2.10	_____	1.25	_____
5	6.3	1.26	4.00	0.80	2.00	0.40
10	7.6	0.76	5.7	0.57	2.50	0.25
20	9.0	0.45	7.2	0.36	2.86	0.14

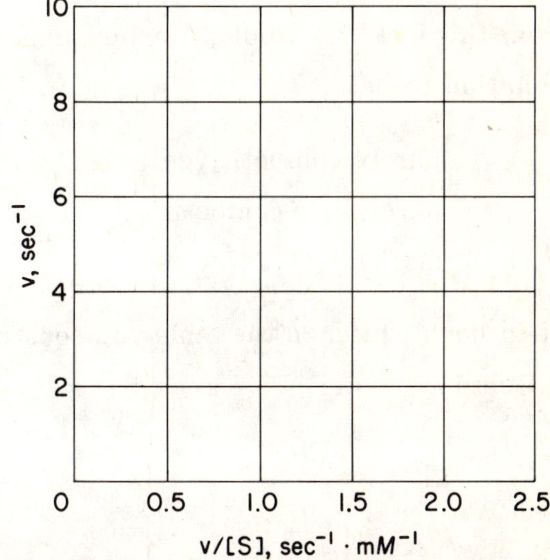

From the results it appears that the line representing the velocity in the presence of a competitive inhibitor and the line representing the uninhibited rate:

( ) will intersect at the ordinate

( ) will intersect below the abscissa

( ) will not intersect

120

Suppose that an inhibitor combines with an enzyme in such a way that only k_2 is decreased. Recall that:

$$K_m = \frac{k_{-1} + k_2}{k_1}$$

and

$$V_{max} = k_2 [E_t]$$

Decide whether, under the steady-state approximation, the result will be a change in:

( ) K_m ( ) V_{max} ( ) both

From this fact, one would, *a priori*, expect the inhibition to be:

( ) purely competitive
( ) purely non-competitive
( ) neither

(If in doubt, refer to the table you constructed in Item 115.)

119

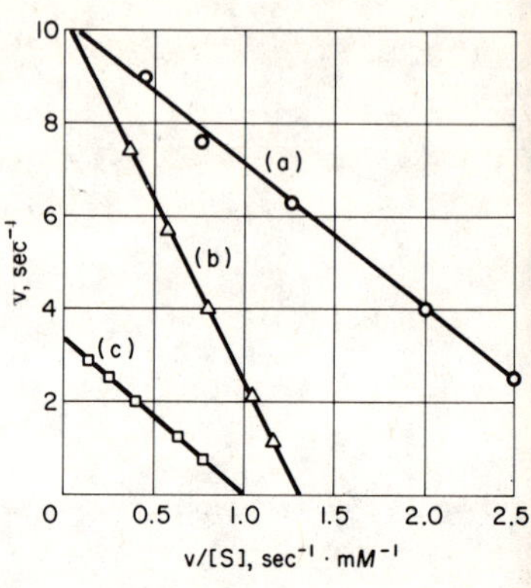

will intersect at the ordinate

120

both

neither

QUANTITATION OF INHIBITION: MEANING OF K_i

121

(Items 121 to 125 may be omitted if no present need is felt to evaluate competitive inhibition quantitatively.) An equation can be derived equivalent to the Michaelis-Menten equation, in which the concentration of the competitive inhibitor as well as that of the substrate appears:

$$v = \frac{V_{max}}{1 + \frac{K_m}{S}\left(1 + \frac{I}{K_i}\right)}$$

The constant corresponding to K_m, which measures the concentration of the inhibitor required to slow the reaction to half the rate it shows in the absence of the inhibitor (given that the observations are made at very low substrate levels) has been designated ______ in this equation.

122

If the competitive inhibitor undergoes conversion to a product, as was true for mannose in our example above

$$E + I \underset{k_{-1}}{\overset{k_1}{\rightleftharpoons}} EI \xrightarrow{k_2} E + P$$

does K_i necessarily represent the dissociation constant of the enzyme-inhibitor complex? ______ (The K_i of mannose in that case should in fact be identical with the K_m obtained using mannose as the substrate, as in Item 103.)

123

If, on the other hand, the competitive inhibitor undergoes no modification by the enzyme, other than to combine with it as follows

$$E + I \underset{k_{-1}}{\overset{k_1}{\rightleftharpoons}} EI$$

K_i may be taken as the __________ constant of the enzyme-inhibitor complex, because there exists no further reaction to prevent equilibrium from being reached by this reversible reaction.

121

K_i

122

No. (Once again, the equilibrium assumption is not justified, *a priori*.)

123

dissociation

124

Yes. (The slope has been increased by:

$$\frac{K_m}{V_{max}} \cdot \frac{[I]}{K_i})$$

124

In Item 78, we rearranged the Michaelis-Menten equation for describing the Lineweaver-Burk plot thus:

$$\frac{1}{v} = \frac{1}{V_{max}} + \frac{K_m}{V_{max}} \cdot \frac{1}{[S]}$$

The corresponding form of the equation (Item 121) that represents, in addition, the influence of a competitive inhibitor is

$$\frac{1}{v} = \frac{1}{V_{max}} + \frac{K_m}{V_{max}} \left(1 + \frac{[I]}{K_i}\right) \frac{1}{[S]}$$

Could one determine K_i from the extent to which the slope of the line has been increased by the presence of the inhibitor? ______

DIXON PLOT FOR DETERMINING K_i

125

A different plot has been used by Dixon to evaluate K_i directly, namely the plot of $1/v$ versus $[I]$ at two or more substrate concentrations. The following data describe an enzymatic reaction in the presence of a competitive inhibitor. Plot the data on the grid below, extending the abscissa scale if you need to. The point at which the two lines intersect, according to Dixon, will equal $-K_i$. Hence the value of K_i in this case is ______ mM. (See Appendix C for the proof of this method.)

	$[S] = 1$ mM		$[S] = 3$ mM	
$[I]$ mM	v min^{-1}	$\frac{1}{v}$ min	v min^{-1}	$\frac{1}{v}$ min
1	1.82	0.55	4.00	0.25
2	1.43	0.70	3.33	0.30
5	0.87	1.15	2.22	0.45
10	0.53	1.89	1.43	0.70

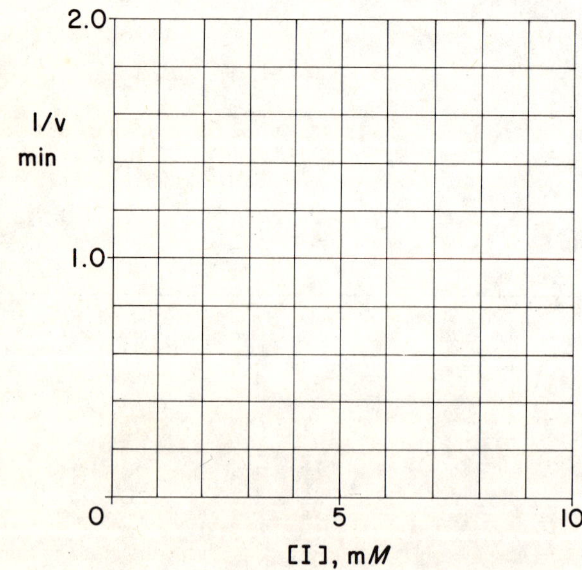

125

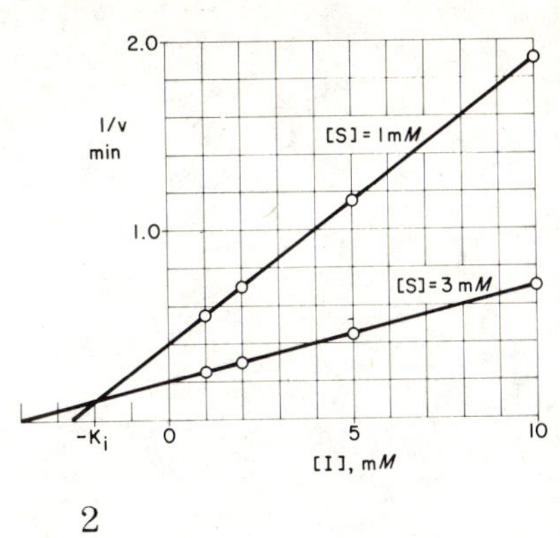

2

KINETICS AND BIOLOGICAL REGULATION; ALLOSTERIC PHENOMENA

126

Many antibiotics and some other drugs produce their biological effects by acting as enzyme inhibitors. Their selection, although still usually made on empirical grounds, depends on their inhibiting a metabolic step more critical to an infective cell or virus than to the host, or a metabolic reaction whose slowing causes an improvement in a deranged balance among metabolic processes.

A rational approach to chemotherapy would involve the identification of a strategic enzymatic reaction, and then the synthesis of compounds with a suitable reactivity with the enzyme concerned. In that connection it would be important to compare the values of [V_{max}] [K_i] for the substances under the test.*

*A more general approach to the kinetics of drug action is to consider that a drug reacts with a specific receptor site, R, to form a complex, without any commitment as to function or location of the structure on which this receptor site occurs. Ariëns has defined a parameter, the *intrinsic activity* of the drug, corresponding approximately to the V_{max}, which defines the maximal biological effect produced when all the available receptor sites are occupied by molecules of the drug; and also an *affinity constant* for the drug corresponding to K_m as defined by Michaelis and Menten. This approach permits one to understand how the simultaneous administration of an analog of a drug may either decrease or increase the total biological effect of the drug, depending on whether the analog-receptor complex has a relatively low (even zero) or a high intrinsic activity. (See E.J. Ariëns, *Arch. int. Pharmacodyn, 99*:32-49, 1954.)

127

Instead of combining at the same active site as the substrate, an inhibitor might combine at a different site, and yet nevertheless modify the reactivity of the substrate, say by slowing the over-all reaction. Thus we would have a ternary complex formed; either

$$EI + S \rightleftharpoons EIS \longrightarrow EI + P$$

or

$$ES + I \rightleftharpoons EIS \longrightarrow EI + P$$

In this situation we probably would do well to call *I* a *modifier* rather than an *inhibitor*, because instead of slowing the over-all reaction it might conceivably __________ it. Such actions are now receiving great attention.

126

K_i

128

It is now well known that living organisms are able to control the various steps of their metabolism by using certain key metabolites as inhibitors or modifiers of enzyme action. A metabolite can, for example, control its own proper concentration in the cell by acting as an inhibitor of the enzyme that participates in its formation. This type of regulation is designated *feedback inhibition* because one can establish a [direct] [reciprocal] relation between the concentration of the product and the activity of the enzyme which it modifies.

127

accelerate

128

reciprocal

129

Usually, it is the final product of a series of reactions, in which a number of enzymes participate, which inhibits the first enzyme in the series, as in the following example:

$$\begin{array}{c} \text{feedback inhibition (F} \dashrightarrow \text{Enz I)} \\ A \xrightarrow{\text{Enz I}} B \xrightarrow{\text{Enz II}} C \xrightarrow{\text{Enz III}} D \xrightarrow{\text{Enz IV}} F \end{array}$$

In this series, the accumulation of F in large amounts will inhibit Enzyme I, thus slowing the series of reactions until the excessive quantity of F is metabolized and the concentration of F is again returned to a level at which its inhibitory action on I is negligible.

This mechanism of control, namely by ____________________, is one that requires a | minimal | maximal | number of interventions by the modifier for the regulation of a series of reactions.

A number of cases are known in which the substrate itself, rather than another metabolite, slows or accelerates the action of the enzyme.

So far we have examined enzyme action in cases in which the activity of the enzyme graphed as a function of substrate concentration takes the form of rectangular hyperbola. Let us now consider a different situation. Plot the following results with care on the graph below.

Substrate concentration (unit unspecified)	% of maximal velocity
5	6
10	16
20	38
30	58
40	76
50	90
100	97

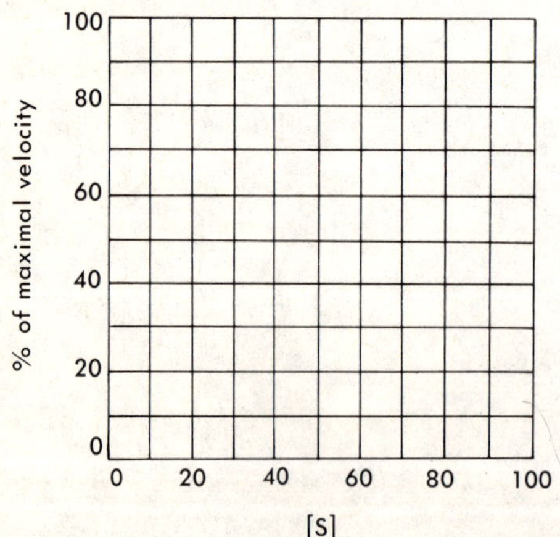

feedback inhibition

minimal

130

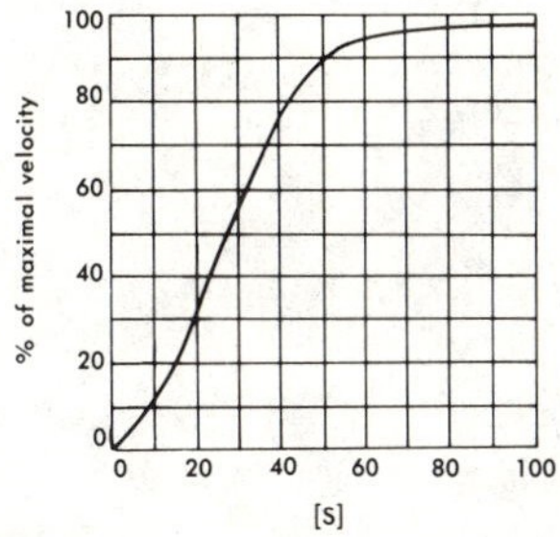

131

On careful examination, one will see that the curve obtained by no means has the form of a rectangular hyperbola. At low concentrations of the substrate, the velocity of the reaction increases relatively slowly; but at levels of the substrate about equal to K_m, one sees that the velocity rises much | less | more | rapidly than it should according to the Michaelis-Menten equation.

131

more

132

In this latter case, the substrate plays a double role. Not only does it react with the enzyme in the usual way (i.e., in the way we have studied in the preceding pages of this program) ; but in addition, the binding of the first molecule of substrate on the enzyme produces a change in the activity of the enzyme. This change may cause a decrease in the value of the K_m for the binding of subsequent molecules of the substrate, and as a result produces | an acceleration | a slowing | of the enzymatic reaction.

133

Usually, the enzymes that show this kind of behavior have more than one active site. We suppose that in the absence of the substrate, these enzymatic sites are intrinsically identical. However, the binding of the first substrate molecule at any one of the active sites modifies the properties of the other active sites of the same molecule of the enzyme. This behavior illustrates the *allosteric* phenomenon.* Because the substrate itself is the modifying agent, this form of the allosteric phenomenon is called *homotropic allosterism.*

In general, each active site of an enzyme molecule showing this behavior is situated on a separate polypeptide chain, a situation implying that the enzyme is composed of two or more identical subunits, each subunit possessing catalytic activity. The isolated subunits show the usual Michaelis-Menten kinetics. When the subunits are assembled to form the original intact molecule, however, the conformational changes which occur when one subunit binds the first substrate molecule are transmitted to the other subunits, with consequent modification of the active sites of the latter.

Thus, at low substrate concentrations, the reaction takes place essentially through the intermediation of a single active site of the enzyme molecule. Under these conditions, since the value of K_m of this site is much [greater | smaller] than when one of the other active sites is already occupied, the rate of the reaction at these low substrate levels is relatively [high | low].

* See bibliographic reference 10.

132

an acceleration

133

greater
low

134

When the substrate concentration is high enough so that the enzyme reaction occurs essentially through the intermediation of the active sites which have lower values for K_m, then the rate of the reaction will be relatively | high | low | . Thus, the allosteric properties of such an enzyme permit it to respond to the presence of an elevated substrate concentration with | a decreased | an increased | activity. This behavior tends to decrease considerably the range of variation in the | concentration of the substrate | the rate of conversion of substrate to product | .

134

high
an increased
concentration of the substrate

135

The relation between the rate of the reaction and the substrate concentration takes the form of a sigmoidal curve, as illustrated in Item 130 , whenever the substrate has a(n) ________________ effect on the enzyme. It has, accordingly, become habitual to suppose that sigmoidal kinetic curves establish the allosteric phenomenon. Because of the complexities presented by the kinetics of two-substrate enzymes (and in fact most enzymes are two-substrate enzymes) it is prudent, however, to recall that the sigmoidal form of the curve may have other origins.

To restate the relation, the sigmoidal form of the velocity-substrate concentration curve suggests but does not prove that the substrate functions not only as a substrate but also as an allosteric affector, i.e., one which modifies the ____________ of the protein molecule and the spatial relations among the ____________ of the protein.

136

So far, a multitude of intermediary metabolites have been observed to act as inhibitors or activators of enzymes. Undoubtedly, some of these effects play a regulatory role under physiological conditions; but it is equally probable that some others do not play such a part. (Among the many examples observed for the actions of hormones on enzymes, most of these cannot represent the normal means of endocrine control.)

Whether or not metabolic intermediates actually serve as physiological regulators will depend on

- ☐ 1. Whether the enzymatic reaction it slows or accelerates is ever rate-limiting to a physiological process.
- ☐ 2. Whether the concentration of the modifier actually varies in the appropriate range under physiological conditions.

135

allosteric
conformation
subunits

136

(Both responses are correct.)

EXPERIMENTAL VISUALIZATION OF *ES*

137

* An event that has emphasized perhaps more than any other the reality of the formation of the enzyme-inhibitor complex, is the observation by Johnson and Phillips* of the high-resolution x-ray diffraction pattern of crystalline lysozyme, and in comparison the x-ray diffraction pattern of the same crystals after exposure to a solution of the competitive inhibitor, N-acetylglucosamine. The pattern of the treated crystals clearly shows a prominence apparently corresponding to the inhibitor molecule at a characteristic crypt or hollow in the surface of the enzyme. Closely similar molecules which are not inhibitory do not cause this prominence. We may expect that the portion of the enzyme which is known as the __________ __________ lies at this locus.

137

reactive site,
active site, or
active center
(*See Item* 9.)

*See Reference 9.

KINETICS OF TRANSPORT

138

* In recent years Michaelis-Menten kinetics have been shown to describe very satisfactorily another type of biocatalysis, namely *mediated transport.*† We may diagram the simplest form of the mediation of transport to show that an *active site,* with which a substrate can form a complex, is located in a membrane in such a way that it can receive the solute from one side of the membrane, and release it unchanged to the other:

$$S + \vdots\; E \rightleftharpoons SE \rightleftharpoons E \;\vdots + S$$

within substance of membrane

(The abbreviation E is used again here without any implication that the active site is located on an enzyme molecule.)

The transport process diagrammed above is one whose operation is symmetrical and fully [reversible | irreversible]. By introducing isotopically labeled substrate (S^*) on either side of the membrane, one can observe that its rate of appearance on the other side obeys the simple Michaelis-Menten equation, providing that one concludes the observations:

( ) before significant amounts of S^* accumulate on the opposite side

( ) at the steady state

†Occasionally one encounters the statement that transport shows "enzyme-like" kinetics. Actually, the kinetic behavior described in this program is by no means specific to enzymatic reactions, and the observation that transport shows that behavior has no force in suggesting enzyme involvement in transport. Had transport kinetics been investigated earlier, we might now be expressing surprise that *enzymes* show *transport* kinetics.

138

reversible

before significant amounts of S^* accumulate on the opposite side

139

* Suppose that the process, as stated, is symmetrical, so that the K_m and V_{max} for transport from left to right are just the same as the K_m and V_{max} for transport from right to left. At a steady state, the uncharged solute will tend to become:

( ) distributed uniformly

( ) concentrated to one side of the membrane

From this behavior, one might suppose that the solute had distributed itself by simple diffusion. Would that conclusion be valid in view of the correspondence of the rate to the Michaelis-Menten equation? ______ That is, should diffusion be subject to saturation by an excess of solute? ______

140

For this system to operate instead to concentrate the solute into the right-hand phase, V_{max} being the same in both directions, which K_m should be the larger?

( ) that for transport from left to right

( ) that for transport from right to left

(Think: At which surface will you want the higher concentration, when the two rates are equal? That surface should show the higher K_m in reacting with the solute.)

Having established such a difference in K_m, an investigator concludes without further evidence that he has proved that uphill transport is produced by a chemical modification in E, such that a lower-affinity form of it is presented on the right-hand side of the membrane and a higher-affinity form on the left-hand side. Does this conclusion follow from acceptance of the steady-state approximation?

______

---

*It should be understood that the formulation

$$S + E \underset{k_{-1}}{\overset{k_1}{\rightleftarrows}} ES \xrightarrow{k_2} E + P$$

either for enzymatic or transport processes is apt to be an abbreviated representation of the full course of events. A longer formulation

$$S + \left| \overbrace{E \underset{k_{-1}}{\overset{k_1}{\rightleftarrows}} ES \overset{k_2}{\rightleftarrows} E'S \underset{k_{-3}}{\overset{k_3}{\rightleftarrows}} E'}^{\textit{membrane}} \right| + S$$

may be desirable at this point to understand the possibility that the affinity of the carrier for the solute is changed on its reorientation from one surface of the membrane to the other.

139

distributed uniformly

No. (Simple diffusion follows first-order or linear kinetics, $v = k \cdot [S]$, at all concentrations.)

No. (A saturating tendency has been claimed possible under extreme conditions.)

140

that for transport from right to left

No.* (It accepts the equilibrium assumption, i.e., that K_m measures the affinity of E for S. True, V_{max} is the same in both directions but this does not establish that k_2 is the same for both directions, unless $[E_t]$ is also identical at both surfaces of the membrane.)

WAYS OF DESCRIBING BINDING BY PROTEINS. GENERAL CONSIDERATION.

141

The Michaelis-Menten equation is based on the proposition that the rate of an enzyme reaction depends only on the concentration of the enzyme-substrate complex. The underlying rate equation is

$$\_\_\_\_\_\_\_\_ = k_2\ [ES]$$

With this equation we use the rate of the enzyme reaction to evaluate the binding of the substrate molecule by the enzyme.

Enzyme kinetics as considered so far in this program may therefore be seen as only a special approach to the study of binding of small molecules by proteins. In enzyme kinetics we measure the | rate | extent | of the catalyzed reaction to evaluate the extent of binding.

141

$v = k_2\ [ES]$

rate

142

The extent of binding may be measured in several other ways. We mentioned in Item 8 that binding may lead to a change in color; and later we spoke of the experimental visualization of ES by x-ray diffraction. Methods based on such changes are often used to measure binding by non-enzymatic proteins.

We have also emphasized that the selective binding of small molecules is a characteristic property of | catalytic proteins only | globular proteins in general |.

Hence our study of kinetics has given us an efficient introduction to the study of binding, which we can briefly develop here for other needs.

For illustration, one general method for measuring binding will be considered here, namely *equilibrium dialysis*. In this procedure, we place the protein in a dialysis bag, and immerse the bag in a buffered solution of the substance to be bound, in the reaction $P + L \rightleftharpoons PL$. This substance, in line with biochemical convention, we will call *the ligand,** L.

The ligand diffuses into the sac until its *free* concentration *inside* and *outside* the sac are the same. In the meantime, molecules of the ligand will also bind to the protein, to an extent described by the equilibrium constant. If we designate the free ligand as L_f and the total ligand in the dialysis sac as L_d, then we can solve for the concentration of the bound ligand,

$$[PL] = [\_\_\_\_\_] - [\_\_\_\_\_]$$

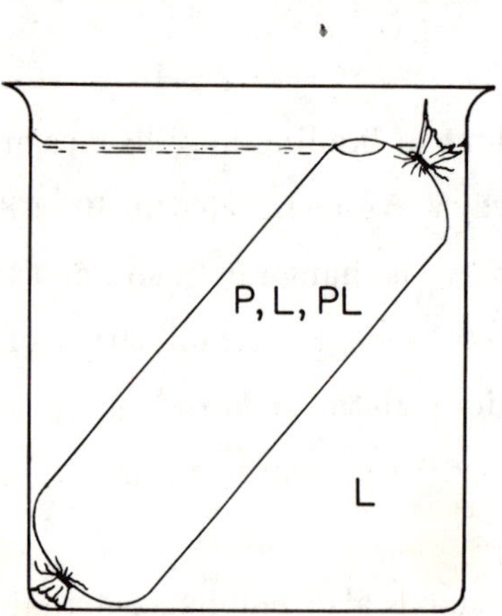

* For obscure reasons, biochemists choose to call the species bound the ligand. Other chemists ordinarily use the term ligand for the binding substance rather than for the substance bound.

globular proteins
in general

143

(The full equation)
$[PL] = [L_d] - [L_f]$

144

To obtain the equilibrium constant for the reaction, $P + L \rightleftharpoons PL$, in the simplest way we will need to measure directly the following concentrations: (Check all answers that apply.)

- ☐ 1. that of the ligand in the external fluid, $[L_f]$
- ☐ 2. that of the total ligand inside the dialysis sac, $[L_d]$
- ☐ 3. that of the bound ligand, $[PL]$
- ☐ 4. that of the total protein, $[P_t]$
- ☐ 5. that of the ligand protein complex, $[PL]$

144

- ☑ 1.
- ☑ 2.
- ☐ 3. (We have already obtained $[PL]$ as equal to $[L_d] - [L_f]$
- ☑ 4.
- ☐ 5. (same comment as for 3.)

145

Thus we come to the general statement of the equilibrium constant, expressed here as an association constant, K_a:

$$P + L \rightleftharpoons PL$$

$$K_a = \frac{[PL]_e}{[P]_e \cdot [L]_e} = \frac{[L_d] - [L_f]}{([P_t] - [L_d] + [L_f]) \cdot [L_f]}$$

Restate algebraically the way in which the three measured quantities are used, this time to obtain the dissociation constant:

$$K_d = \frac{\qquad}{\qquad}$$

146

Note also for a simple binding process that K_a will be expressed in | molar | molar^{-1} | units, and K_d in ________________ units. Historical precedent favors the use of the form K_d in binding studies.

145

$$K_d = \frac{[P]_e \times [L]_e}{[PL]_e} =$$

$$\frac{([P_t] - [L_d] + [L_f]) \cdot [L_f]}{[L_d] - [L_f]}$$

(The full equation)

147

Now let us introduce a term, the *fractional saturation* of P, using the abbreviation $\bar{v}$, defined for the case where there is only one site per protein molecule*:

$$\bar{v} = \frac{\text{total moles of substrate bound}}{\text{total number of sites available, filled and unfilled}}$$

Note the analogy to the *relative velocity,* v/V_{max} footnote to Item 88.

If a given protein has only a single binding site for L, then as $[L]$ increases from zero to infinity, $\bar{v}$ will increase from zero to __________.

*A popular alternative definition sets v equal to moles of ligand bound/moles of protein.

146

molar^{-1}

molar

147

one (unity)

148

Abbreviating, we may rewrite our definition of the fractional ______________:

$$\bar{v} = \frac{[PL]_e}{[P]_e + [PL]_e} = \frac{[PL]_e}{[P_o]}$$

where P_o stands for the total concentration of protein.

Since

$$K_a = \frac{[PL]_e}{[P]_e \cdot [L]_e}$$

and since $[P]_e = [P_o] - [PL]_e$

$$K_a = \frac{[PL]_e}{[P_o]\ [L]_e - [PL]_e\ [L]_e}$$

If we rearrange in steps

$$K_a \cdot [P_o]\ [L]_e - K_a\ [PL]_e\ [L]_e = [PL]_e$$

$$K_a\ [L]_e \cdot [P_o] = (1 + K_a\ [L]_e)\ [PL]_e$$

we obtain

$$\frac{K_a\ [L]_e}{1 + K_a\ [L]_e} = \boxed{} = \bar{v}$$

149

Under binding studies, we may omit the subscript e if we agree that we are discussing binding at equilibrium.

Since $K_a = \frac{1}{K_d}$, the equation $\overline{v} = \frac{K_a [L]}{1 + K_a [L]}$

can be written in the form $\overline{v} = \frac{1}{1 + \frac{K_d}{[L]}}$

Note the parallelism of this equation to the two forms of the Michaelis-Menten equation (Item 44). What graphic form does the relation between $\overline{v}$ and $[L]$ describe,

| a straight line | a rectangular hyperbola | a sigmoid curve | a parabola |?

148

(The full equation)

$$\frac{K_a [L]_e}{1 + K_a [L]_e} = \frac{[PL]_e}{[P_o]} = \overline{v}$$

149

a rectangular hyperbola

150

In the accompanying figure, the asymptote of the curve shows the number of ________________ per molecule of protein.

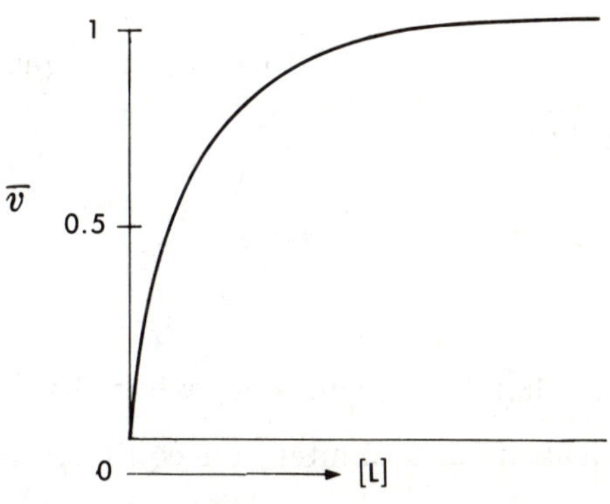

By inspection of the equation $\overline{v} = \frac{1}{1 + \frac{K_d}{[L]}}$

What value should the concentration of the ligand take (in algebraic terms) to produce 50 per cent saturation of the protein? ________________

You will recognize that the Michaelis-Menten equation

$$v = \frac{V_{max} \cdot [S]}{K_m + [S]}, \text{ or } \frac{v}{V_{max}} = \frac{1}{1 + \frac{K_m}{[S]}}$$

is merely a special form of this general binding equation

$$\bar{v} = \frac{1}{1 + \frac{K_d}{[L]}}$$

and that K_m becomes K_d when, as postulated by Michaelis and Menten, an equilibrium of binding can occur.

We may say that there is only one important equation in the range of binding and enzyme kinetics, and that is the equation for a rectangular hyperbola. The literature is full of equations, all variants of this one, but each bearing a different name.

We have already seen how the binding equation can be manipulated to give linear forms rather than a hyperbolic plot. Another common alternative is probably known to you from your study of pH and dissociation, namely the Henderson-Hasselbalch equation.

Because the association constant is given by *

$$K_a = \frac{[PL]}{[P]} \cdot \frac{1}{[L]}$$

* Recall that these concentrations refer to the condition of equilibrium, but that we have agreed to omit the subscript *e*.

binding sites (or equivalent words)

$[L] = K_d$ (In words, the protein will become half-saturated when the concentration of the ligand equals the dissociation constant.)

then

$$K_a \, [L] = \frac{[PL]}{[P]}$$

or in logarithmic form, $\log K_a + log\ [L] = log\frac{[PL]}{[P]}$

In analogy to the symbol pH for $-\log[H^+]$ we may agree to state the concentration $[L]$ in the form of its negative log

$$p[L] = -\log\ [L]$$

then our logarithmic equation can be rewritten

$$-\log K_a - \log L = -log\frac{[PL]}{[P]}$$

$$\log K_d - \log L = \log\frac{[P]}{[PL]}$$

$$p[L] = pK_d + \log\frac{[P]}{[PL]}$$

$$\boxed{} + \log\frac{[A^-]}{[HA]}$$

Complete in the box the usual form of the Henderson-Hasselbalch equation to show the identity of the two forms.

152

We can write this equation also in terms of the ratio between the unfilled and filled sites on the binding protein. Since $\bar{v}$ is the proportion of the total sites that are filled

$$\bar{v} = \frac{[PL]}{[P_o]}$$

then $1 - \bar{v}$ is the proportion of the total sites that are unfilled

$$1 - \bar{v} = \frac{[P]}{[P_o]}$$

In correspondence, the ratio of unfilled sites can be stated as

$$\frac{1 - \bar{v}}{\bar{v}} = \frac{[P]}{[PL]}$$

Write the logarithmic form of the binding equation (Item *151*) in terms of the ratio of unfilled to filled sites:

151

(The full equation)

$$pH = pK + \log \frac{[A^-]}{[HA]}$$

(pK may also read pK' or pK_a')

152

(The full equation)

$$p[L] = pK_d + \log \frac{1 - \bar{v}}{\bar{v}}$$

153

This logarithmic form of the binding equation is often related to a plot of $1-\bar{v}$ (the fraction of sites empty) against $-\log[L]$. Such plots are no doubt familiar to you as the plots of the fractional dissociation of a weak acid as a function of pH.

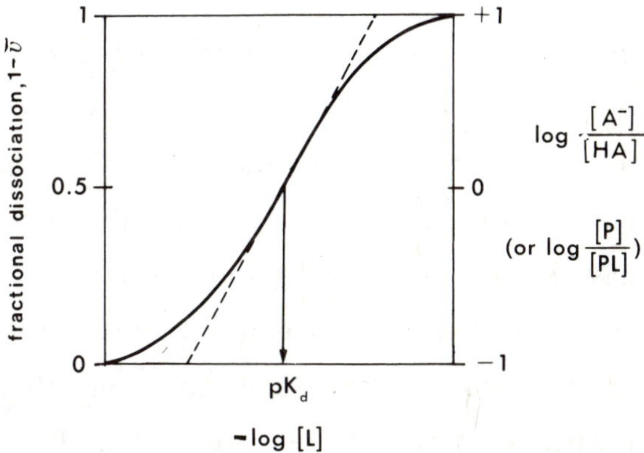

This correspondence allows you to extend all you have learned under the heading of pH and dissociation in using this curve to describe the binding of species other than the hydrogen ion. For example, you will recall that the curve for any given dissociation takes a characteristic position, as measured by its midpoint, with respect to concentration of L. At this midpoint $\bar{v}$ equals $1 - \bar{v}$ (and $[P]$ equals $[L]$) hence, in the equation;

$$p[L] = pK_d + \log \frac{1 - \bar{v}}{\bar{v}}$$

the term $\log \frac{1 - \bar{v}}{\bar{v}}$ has a value of zero. The equation tells us then that half-saturation will occur when the ligand concentration is equal to________

__________________.

154

You may perhaps recall that the slope of the midpoint of the sigmoidal curve is always the same for the simple titration of a monobasic weak acid. We can evaluate this slope better by plotting log $\frac{1 - \bar{v}}{\bar{v}}$ (or log $[P]/[PL]$) against log $[L]$, which gives us a straight line instead of a sigmoidal curve. (The dashed line in Item 153, which refers to the scale at the right, shows how this mode of plotting straightens the sigmoid.) The slope then has a constant value of 1 for any reaction in which only one ion or molecule enters a binding reaction, or in which each site shows an unmodifed K_{diss} regardless of the per cent saturation. This slope is known as the *Hill coefficient*. Evaluation of the Hill coefficient is very important for the study of binding such as that shown by hemoglobin. Here the binding of the first O_2 molecule modifies that of the other O_2 molecules that bind subsequently. The result is a slope with a value other than one. (In the case shown in Item 130, $\bar{n} = 2.8$).

The allosteric phenomenon, as discussed in Items 128 to 134, | should | would not, however, | lead to non-integral values of $\bar{n}$.

153

K_d
half-dissociated

154

should

155

$\frac{K_d}{[P_o]}$

$\frac{1}{[P_o]}$

155

We may turn now from the two curvilinear plots of Items 150 and 153 to two linear transformations of the binding equation. The first of these corresponds to the Lineweaver-Burk plot. The binding equation (Item 149) is arranged to place $[L]$ and $[PL]$ in their reciprocal forms.

$$\frac{1}{[PL]} = \frac{1}{[P_o]} + \frac{K_d}{[P_o]} \cdot \frac{1}{[L]}$$

This equation, known as the Benesi-Hildebrand equation, is used extensively in chemistry. If one plots $\frac{1}{[PL]}$ against $\frac{1}{[L]}$, one obtains a straight line whose slope is ____________, and whose intercept on the Y axis is ____________.

156

The slope divided by the intercept will accordingly yield the value for __________. Or, instead, one can obtain this value by extrapolating the line to the abscissa, just as one obtains the value of __________ in the Lineweaver-Burk plot.

157

Another famous plot used in the study of binding is the Scatchard plot. It is based on the binding equation in the form

$$\frac{\bar{v}}{[L]} = K_a\ (1 - \bar{v})$$

One plots $\frac{\bar{v}}{[L]}$ against $\bar{v}$. The analogy to the Hofstee or Augustinsson plot should be noted, although conventionally the axes are interchanged from the orientation used in that kinetic plot.

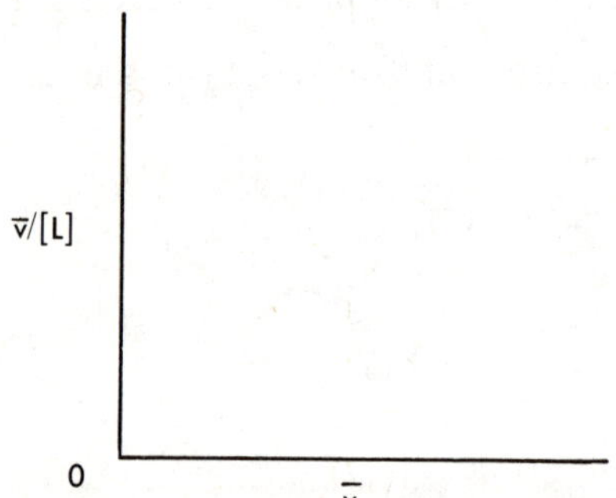

From inspection of the equation we can see that the slope of the line is $-K_a$. The intercept on the abscissa will be 1 if there is only one binding site per molecule of the binding agent. That is, this intercept shows:

☐ 1. The degree of saturation of each binding site.

☐ 2. The number of binding sites per molecule of the binding agent.

156

K_d

$1/\ K_m$

157

The number of binding sites per molecule of the binding agent.

158

Because all of these plots are based on variants of the same equation, none is intrinsically superior. Each has advantages in given situations, and they are used interchangeably.

Perhaps the rarest usage, even though it is quite satisfactory, is as illustrated, the plotting of the logarithmic forms of the Michaelis-Menten equation. In such plots the position of the midpoint of the sigmoidal curve, as projected in the diagram, will correspond to

$-\log K_m$	$-\log V_{max}$

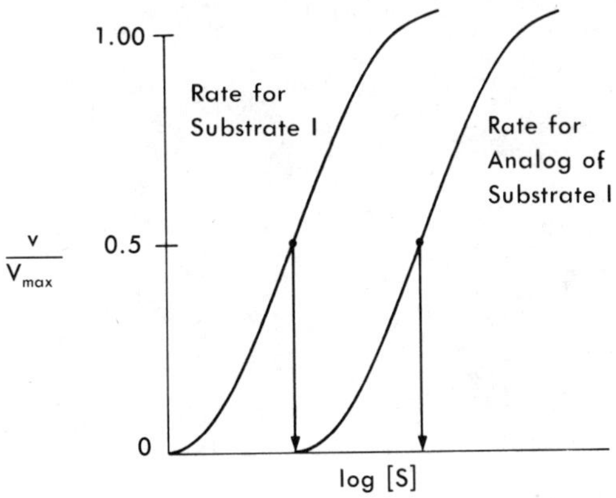

1.00
Rate for Substrate I
Rate for Analog of Substrate I
$\frac{v}{V_{max}}$
0.5
0
log [S]

159

Restate here again the proposition that permits the same equation for a rectangular hyperbola to describe both

a) the binding of a ligand by another substance, and

b) the rate of a simple enzymatic reaction,

each as a function of the ligand or substrate concentration.

158

$-\log K_m$

159

The rate of the enzymatic equation depends on the amount of the enzyme-substrate complex. Hence the rate equation is merely a special form of the general binding equation (or words to that effect).

EFFECT OF TEMPERATURE

160

* Two other factors, besides substrate concentration, are important in controlling the rate of enzymatic reactions, namely *temperature* and *pH.* A study of these factors is, however, considerably more difficult, both technically and theoretically, than the study of the effect of concentration; hence we will only touch on the phenomena involved.

As for other chemical reactions, the velocity of an enzymatically catalyzed reaction increases with an increase in temperature. Above a certain critical temperature, however —and this critical temperature varies from enzyme to enzyme—the reaction velocity begins to *decrease* and at high temperatures the enzyme becomes completely inactive. From the kinetic standpoint this decrease is of minor interest in that it merely reflects the inactivation (usually through denaturation) of the protein molecule. In the figure below, mark the portion of the activity curve that describes the behavior of the enzymatic reaction shared with other chemical reactions, and the portion that arises from denaturation.

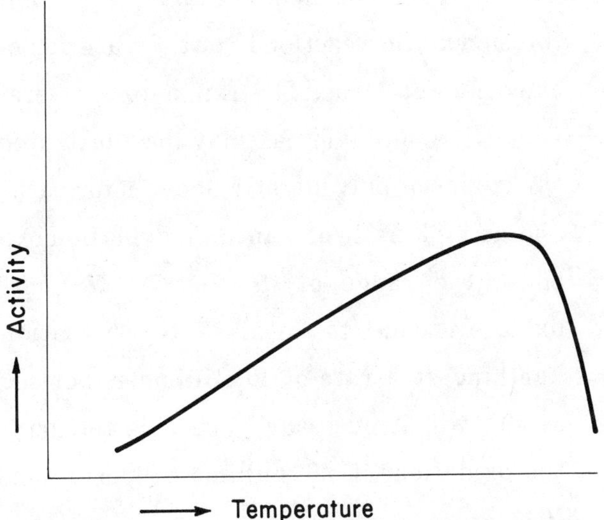

Page 111

161

* The increase of enzymatic activity at moderate temperatures is much more interesting to us at this point. It is a consequence of the transfer of thermal energy to the various bond-making and bond-breaking processes that occur during the reaction. A careful study of the effect of temperature may reveal the heat change, the free energy change, and the entropy change associated with a given step in an enzymatic reaction.

For biological systems, the parameter used most frequently to describe the rate increase with temperature is the $Q_{10°}$, which represents the number of times by which the enzymatic activity is increased by a 10° rise in the temperature. If, for example, an enzyme shows a V_{max} of 400 min^{-1} at 27° and a V_{max} of 800 min^{-1} at 37°, the $Q_{10°}$ over this interval is ______.

160

Activity

Shared with other Chemical Reactions

Arises from Denaturation

Temperature

162

* Values of $Q_{10°}$ of about 2 are very common for enzymatic reactions, as they are for other chemical reactions. The range encountered is very wide, however, as may be illustrated by $Q_{10°}$ values of only slightly above 1 for catalase, and as high as 4 for xanthine oxidase, measured in the range of 27° to 37°. If xanthine oxidase is able to catalyze the oxidation of xanthine at a rate of 10 millimoles per second at 40°, will it necessarily be able to carry out the oxidation at 40 millimoles per second at 50°? ______

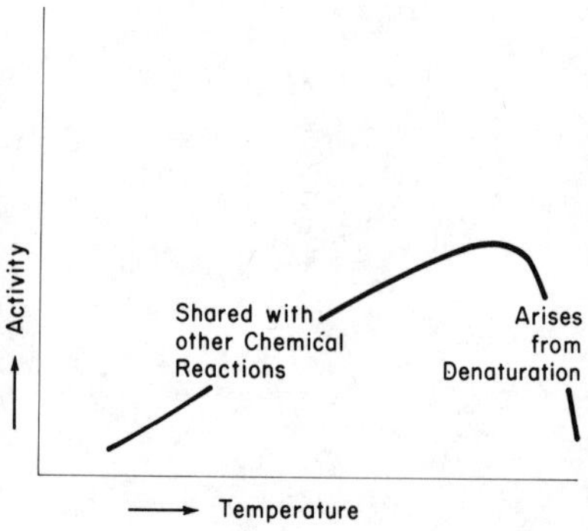

161

2

162

No. (The temperature of 50° may well cause extensive loss of activity by denaturation. In general a $Q_{10°}$ value determined under one set of conditions cannot be applied indiscriminately under others.)

163

* The figure below represents the difference between the energy content of the reactant and the product in a simple chemical reaction

$$A \underset{k_{-1}}{\overset{k_1}{\rightleftharpoons}} B$$

(An activated state of A, namely A^*, lies in the pathway taken by the reaction.)

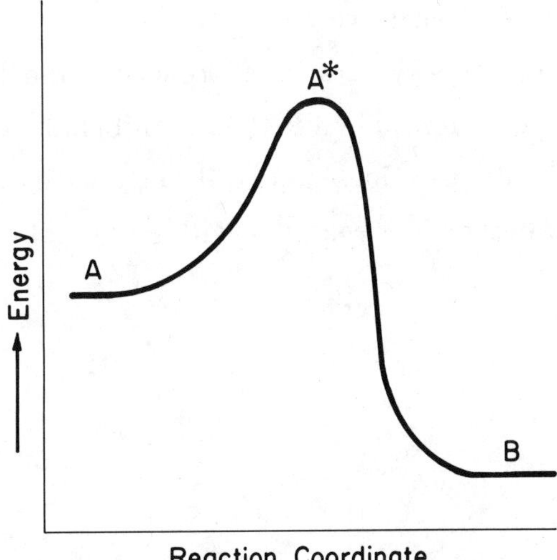

Because B is at a lower energy level than A, the equilibrium for the reaction will favor the formation of B. That is, the equilibrium constant may call for a concentration of B perhaps:

( ) 10 or 20 times that of A

( ) one-tenth or one-twentieth that of A

164

* For A to be converted to B, however, it must pass through the activated state A^*, which is energetically an upward course. If the "energy hump" represented by A^* (the *activation energy*) is small, then A will be changed easily to B and k_1 will be large. If, however, the energy of activation is large, not enough energy may be available in the system to surmount the energy hump, and k_1 may be zero. Hence we see that the total change in free energy in a given reaction (and therefore the magnitude of the equilibrium constant) | tells us | does not tell us | with certainty whether the reaction will go or not.

163

10 or 20 times that of A

164

does not tell us

165

There are many such cases in which the equilibrium constant favors the occurrence of a reaction, but the size of the activation energy prevents it. Thus the mixture of O_2 and N_2 in the atmosphere is *thermodynamically unstable,* since the magnitude of the equilibrium constant predicts that they will combine. Because the energy of activation needed for them to react is very great, however, the mixture is *kinetically stable.* That is, the large activation hump causes the rate constant for reaction to be exceedingly small.

We may draw the generalization that when the energy difference $(A^* - A)$ is relatively small, k_1 will be | large small |; and vice versa. By the same generalization, if the energy difference, $(A^* - B)$ as illustrated, is larger than $(A^* - A)$, k_{-1} will be | larger smaller | than k_1.

166

* This concept is readily carried over to enzymatic reactions, although the energy profile is usually more complex, as illustrated in the figure below. The enzyme operates to open a

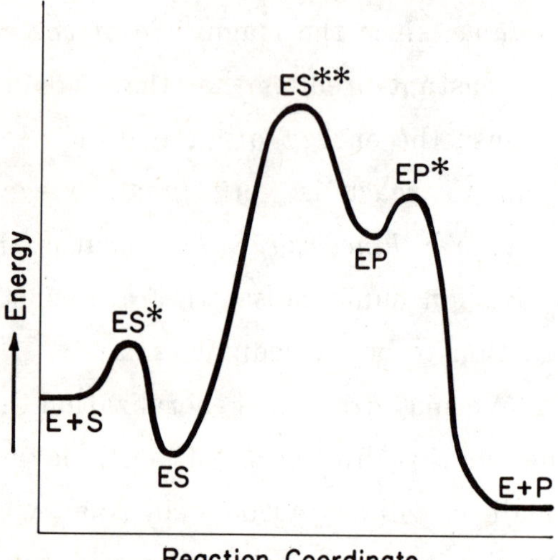

route from substrate to product that avoids any activation energy so high that k_1 is negligibly small; hence a reaction that otherwise proceeds only at a negligible rate will now proceed spontaneously and rapidly. The case illustrated involves | one | two | three | activation steps. The over-all energy profile favors the conversion of | S to P | P to S |. Despite the simplicity of the concept, the complexities of identifying the various contributions to the energy profile are well beyond the present treatment. (Interested students should turn to Reference 4.)

165

large

smaller

166

three

S to P

EFFECT OF pH

167

* The effect of pH on the velocity of an enzymatically catalyzed reaction is frequently pictured ideally as illustrated below. The idealized curve is bell-shaped, formed from two components, shown here by the solid lines, which presumably represent the dissociation curves of two groups important to catalysis of the reaction.

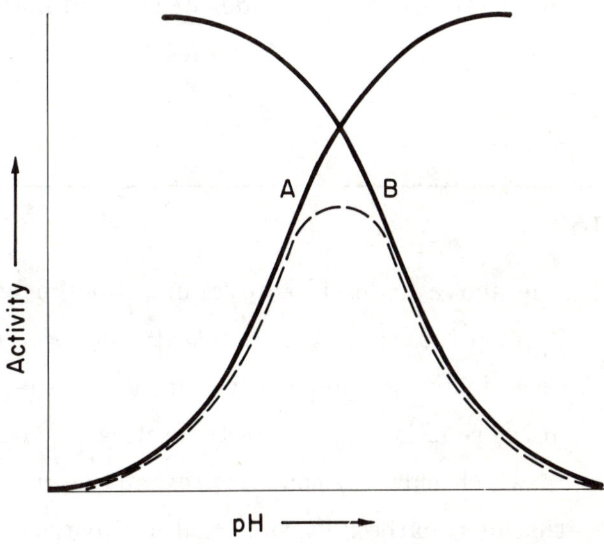

Suppose that the figure represents the case of an enzyme containing at the active site a glutamyl residue and a histidyl residue. Furthermore let us suppose that this enzyme will be effective only if the γ-carboxyl group of the glutamyl residue is in the negatively charged (deprotonated) form, and the imidazole group of the histidyl residue is in the positively charged (or | protonated | deprotonated |) form. In the figure above, curve *A* shows how the proportion of the enzyme molecules having this carboxyl group in the effective form increases as the pH is raised. As the pH is further increased, however, the proportion of the imidazole groups in the effective form will | increase | decrease |, as described by curve *B*. The composite of these two effects is approximated by the dashed line, forming a bell shaped curve.

168

Theoretically, should the midpoints of the two curves forming the left and right sides of the bell-shaped curve in Item 167 correspond roughly to the pK′ values of the two hypothetical dissociating groups critical to the reaction? ______ Picture mentally how the bell-shaped curve would look if the two pK′ values were not separated at all. Would the midpoints of the two sides of the bell indicate accurately the two pK′ values? ______

167

protonated

decrease

169

* The above example is oversimplified because it requires that the substrate not have a dissociating group that will influence the rate of the reaction. Cases are theoretically possible in which curve *A* could represent the dissociation of a carboxyl group on the substrate and curve *B* the dissociation of an imidazole group on the enzyme. Our example also assumes that no other dissociable group on the enzyme has the slightest effect on the activity of the enzyme. We have the further possibility that with the formation of *ES*, a new acidic dissociating group may be formed, differing from those shown by either *E* or *S*. A curve showing how V_{max} changes with pH would tend to describe a hydrogen-ion dissociation:

( ) attributable to the enzyme-substrate complex

( ) attributable to the free enzyme

168

Yes

No. (The two pK′ values must be separated by at least two pH units to permit their estimation in so direct a manner.)

169

attributable to the enzyme-substrate complex (Since at V_{max} by definition, no E will be present. The dissociable group concerned may of course also be present on the free enzyme.)

170

* Some efforts have been made to determine, by measuring the change of K_m and V_{max} with pH, whether a given dissociable group participates in binding the substrate, or in the subsequent catalytic event. (Reference 5 may be consulted on this question.) An unambiguous result requires that the quantitative relations among the rate constants k_1, k_{-1}, and k_2 be determined. Ordinary observations of the effect of substrate concentration on the rate of an enzymatic reaction

( ) readily reveal that relationship

( ) unfortunately do not reveal that relationship

170

unfortunately do not reveal that relationship

IMPLICATIONS FOR PRACTICAL ENZYMATIC ANALYSES

171

In his study of kinetics the student is frequently assured that an elementary knowledge of the subject is essential for the reliable design and performance of enzyme assays. Although this proposition may be approximately correct, the student is usually asked to accept it on faith. Here we will stop to collect and note some of the simpler consequences of kinetics for the analytical use of enzymes, and perhaps in that way convince the student that the above proposition is valid. More important, the questions asked should serve for practical review.

Enzyme assays are of two principal kinds, those in which the enzyme serves as a specific reagent for measuring the amount of the substrate present, and those in which the purpose instead is to measure the amount of the enzyme present.

For the first case we usually prefer to cause the reaction to go substantially to completion and then observe how much of a characteristic product is formed. Our first question here is how much enzyme to use, or how long a time to allow when we use a given amount of enzyme. This preliminary question is a kinetic question.

In this case we want to be sure to provide the enzyme

at a rate-limiting level	in excess

171

in excess. (Granted, you were already able to answer this question in Item 2; but now we need to return to this mode of reasoning.)

172

The calculation of the time to allow, or the amount of enzyme to use, when the concentration of the substrate is well below the K_m from the beginning, is a simple one, because then the reaction will effectively be | zero-order | first-order | second-order | with respect to the substrate.

172

first-order

173

Recall that for a first-order reaction we can extract a constant value for a half-time, namely the time required for the reaction to go halfway to completion. Such a half-time can be obtained for any selected concentration of enzyme.

Suppose we are satisfied with a 98 per cent conversion of the substrate to the product. For this purpose, estimate how many half-times we should allow. | three | five or six | over 10 | half-times.

173

five or six. (In one half-time, 50 per cent completion. In two half-times, 70 per cent; in four half-times, 94 per cent; in six half-times, 98.6 per cent completion.)

174

If the initial concentration of the substrate is nearly as large or larger than the K_m, the rate at first will be slower than first-order, and | more | less | time will be needed.

175

174

It is often convenient to convert the product of the first enzymatic reaction to a second product by use of a second enzyme, because the second product is more easily measured. In such *coupled* enzymatic reactions for substrate assays, optimal economy in the quantities of the enzymes needed is obtained by using the enzymes in favorable proportions, so that neither step is far removed from being rate-limiting.

$$S \xrightarrow[\text{present in excess}]{\text{Enzyme I to be}} P_1 \xrightarrow[\text{present in excess}]{\text{Enzyme II to be}} P_2$$

In these cases the calculation becomes more complex, and graphs have been offered for estimating the time to be allowed, or the quantites of the enzymes that should be used to produce the conversion in a given time (see reference 11). In any case, we may safely say that if the first reaction is not significantly reversible, the coupled reaction will take | a shorter | at least as long an | interval of time as the first reaction alone.

more

175

at least as long

176

We have emphasized that it is desirable to have any enzyme used as a reagent present in excess. Which of the following appear to be good reasons for not setting the excess too high?

- ☐ 1. The enzyme is in short supply and expensive to prepare or buy.
- ☐ 2. The enzyme preparation is not pure but contains other enzymes that can come to interfere if too much is used.
- ☐ 3. The enzyme itself can produce an interfering side reaction, which may become conspicuous if the excess is large.

176

(All are considered valid reasons.)

177

Let us turn, then, to the second kind of study, in which our purpose is to measure how much of an enzyme is present in a given sample. This enzyme must be present in [excess | rate-limiting concentrations], whereas we should like to have present in excess [the substrate, alone | all substrates and cofactors]

177

rate-limiting concentrations

all substrates and cofactors

178

We expect to observe under these conditions [the maximal velocity | the maximal relative velocity, v/V_{max}], applying to the concentration selected for the enzyme.

179

This conclusion follows directly from the definition that $V_{max} = k_2[E]_t$. Unless the enzyme is completely converted to the enzyme-substrate complex, we shall not observe the effect of its total concentration.

We must confess that it is possible to obtain a rate *proportional* to the total concentration of the enzyme if we make certain that a fixed proportion of the enzyme (say ⅔ or ¾) has been accurately converted to the complex. Which of the following technical details would be likely to threaten the accuracy of the analysis in that case?

☐ 1. The concentration of the substrate is set with an uncertainty of ± 10 per cent.

☐ 2. An unsuspected competitive inhibitor is present in some of the samples.

178

the maximal velocity

179

(Both details would be threatening; both argue for the advantage of keeping the enzyme saturated with substrate.)

180

It is not always convenient, however, to have all cofactors present in excess. In the reaction

$$S + NADH + H^+ \xrightarrow[\text{is to be determined}]{\text{enzyme whose concentration}} SH_2 + NAD^+$$

we often find it convenient to measure the rate of disappearance of the coenzyme NADH, or, in parallel instances, of NADPH. Here we cannot get good analytical accuracy if we use a large excess of NADH, because then the proportion of it consumed in the course of the assay would be too small for accurate measurement. Hence, we may be forced to use a concentration of NADH well below that required to saturate the enzyme reaction, although *S* should as usual be present in excess. Note also the importance of controlling the pH.

If we consider the theoretical implications of carrying out an enzyme assay under these conditions, we may see that the velocity obtained is | without qualification, the V_{max} | a value of V_{max} applying only for the arbitrarily selected conditions | .

181

Enzyme assays frequently use two-step or coupled enzymatic reactions. The procedure is to use a second enzyme as a reagent to measure how much product the first enzyme generates, as illustrated:

$$S \xrightarrow[\text{rate is being measured}]{\text{Enzyme I, whose}} P_1$$

$$P_1 \xrightarrow[\text{must not determine the result}]{\text{Enzyme II, whose rate}} P_2$$

This choice is usually made when P_2 is more conveniently measured than P_1. For cases like the one illustrated, we want Enzyme ________ present in excess, and Enzyme ________ present at rate-limiting concentrations.

180

a value of V_{max} applying only for the arbitrarily selected conditions (See Item 98 for the procedure for determining the value of V_{max} applying when the enzyme is fully saturated with the cosubstrate.)

182

Because the first step releases product P_1 gradually to the second enzyme, and so on, all steps but the first step in a multi-step enzyme assay will generally be first-order. Under the preferred condition that the substrate(s) be present in excess, the first step will be [zero-order | second-order]. The lag between the formation of P_1 and the formation of P_2, represented by the extent to which P_1 accumulates, will tell us how much extra time to allow when there is a second step. This lag depends solely on the first-order rate constant for the [first | second] step; in fact, it can be shown by a calculation which we will not develop here that the lag will never exceed 1.44 half-times for that step.

181

II } (The case is covered by the principle that all reac-
I } tants and cofactors should be present in excess for an enzyme assay.)

182

zero-order

second

183

The pH's selected may well be different. (In the first case V_{max}/K_m (Item 51) will define the optimal pH; in the second case V_{max} alone will be decisive.)

183

The preceding 12 items have illustrated how the study of kinetics may guide us in the design of analytic procedures that use enzymes. Other factors that should be considered in setting the conditions for an assay include the temperature and the pH. Obviously, one should not select a temperature so high that enzyme breakdown during the assay is a significant factor. Nor should one assume that all enzymatic progress of the reaction can be prevented until the start of the assay by keeping the reaction mixture in an ice bath.

As suggested by Item 170, the pH for maximal velocity of an enzymatic reaction does not necessarily coincide with the pH for a minimal value of K_m, because V_{max} depends on the dissociation of groups on the ES molecules (via k_2), whereas K_m is sensitive to the dissociations of E, S, and ES (via k, k_{-1}, and k_2). Taking this difference into account, decide whether or not the optimum pH selected will necessarily be the same for (a) the assay of the quantity of substrate present, carried out at a low substrate level and (b) an assay for the quantity of enzyme present, carried out at saturating or near-saturating substrate concentration.

SUMMARY

For items requiring lengthy textual answers, please be satisfied with an approximate correspondence of your response to the printed one.

184

* The Michaelis-Menten equation, which is:

$$v = \frac{\qquad}{\qquad}$$

describes the dependence of the rate of an enzymatic reaction on the substrate concentration. It is especially suitable for obtaining the rate at substrate concentrations:

( ) below $K_m/10$

( ) from $K_m/10$ to 10 K_m

( ) above 10 K_m

(*See Item 52!*)

184

$$\frac{V_{max} \cdot [S]}{K_m + [S]}$$

from $K_m/10$ to 10 K_m

185

ES is broken down as fast as it forms

185

The mass-action law tells us that the rate of an enzymatic reaction depends on the concentration of the complex, ES. The most generally applicable assumption that will permit us to proceed on this fact to the Michaelis-Menten equation is the following: During a short period of observation

( ) ES is broken down as fast as it forms

( ) ES passes on to E and P much more slowly than it reverts to E and S

(*Item 36*.)

186

* If that is true, then K_m may be described as having the same value as the concentration of substrate that will cause ____.

Only if the second assumption also applies does the K_m come to have the additional significance that it measures inversely the affinity of the enzyme for the substrate, hence that it is ____.

(*Items 47, 67.*)

187

Defined in words, the term V_{max} means:

____

____

Algebraically it may be defined most simply as follows:

$$V_{max} = \_\_\_\_\_\_\_\_\_\_$$

If it is stated in terms of moles of substrate converted per mole of enzyme per min, V_{max} may also be called the __________ number. (*Items 43, 50.*)

188

* The advantage of rearranging the Michaelis-Menten equation for plotting by the method of Lineweaver and Burk, or that of Augustinsson or of Woolf, is that:

____

____

____

In the first of these methods, one plots ______ against ______. (*Items 79, 92.*)

189

* The presence of a competitive inhibitor does not act to change the [K_m | V_{max}], although an increased value of the ______ is obtained. A non-competitive inhibitor decreases the [K_m | V_{max}], although the ______ is not changed. (*Item 115.*)

186

a half-maximal rate

the dissociation constant of ES (with respect to E and S)

187

the highest velocity that can be obtained (other conditions being constant) by raising the substrate concentration

$k_2[E_t]$ (not $k_2[ES]$, unless $[S]$ is infinitely high)

turnover

188

We obtain a plot whose points theoretically describe a line, the slope and intercepts of which together yield the values K_m and V_{max}.

$1/v$, $1/[S]$

189

V_{max}, K_m

V_{max}, K_m

190

A third kinetic parameter (*beyond* the K_m and the V_{max}) measures the reactivity of an *inhibitor* with the enzyme. This measure is the ______. Like the K_m, it is measured in units of | concentration | rate |. (*Item 121.*)

190

K_i

concentration

191

* What a transport mediator does is to act catalytically to facilitate the transfer of a substance from one phase to another. The most general way one can recognize its intervention is by noting that the substrate:

( ) moves against a concentration gradient

( ) "saturates" its own transport

(*Item 139.*)

191

"saturates" its own transport (*Item 5.*)

(Mediated transports are not always "uphill")

192

* Elevating the temperature tends to have a | similar | different | effect on enzymatic and non-enzymatic reactions, provided that one avoids temperatures high enough to cause __________________ of the enzyme. (*Item 160.*)

193

* The effects of pH on enzymatic reactions follow from the requirement that certain critical groups on the enzyme, on the substrate or on the enzyme-substrate complex must be present in a specified state with regard to ________________. (*Item 167.*)

192

similar

inactivation (denaturation)

194

The Michaelis-Menten equation has the same form as a fundamental equation describing the binding of one molecule by another, as a function of the concentration of the molecule being bound,

$$\bar{v} = \frac{1}{1 + \frac{K_d}{[L]}}$$

because the Michaelis-Menten equation assumes the rate of an enzyme reaction to depend on the extent to which the enzyme binds the substrate.

State in words the meaning given to the three symbols in this binding equation:

$\bar{v}$ ________________

K_d ________________

L ________________

(*Items 148-151*)

193

association or dissociation of H^+ (protonation)

194

$\bar{v}$ = fractional saturation (of the binding agent)

K_d = the dissociation constant of the complex (concentration of ligand needed to produce half-saturation)

L = the ligand, or molecule bound

195

By inserting a, b, c, or d in the blanks below, match these methods for describing binding with the form of the plots typically obtained as illustrated below:

a. The binding equation of the preceding item.

b. The Benesi-Hildebrand equation, $\frac{1}{[PL]}$ graphed against $\frac{1}{[L]}$.

c. The Henderson-Hasselbalch equation.

d. The Scatchard plot, $\frac{\bar{v}}{[L]}$ graphed against $\bar{v}$.

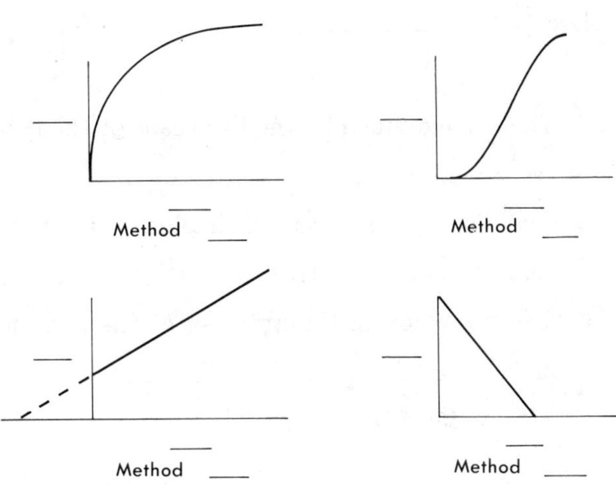

Insert labels on the abscissas and ordinates to show what is plotted in each case.

(*Items 150-157*)

196

Identify the significance of the following features of the four plots by inserting the appropriate symbol in each blank from this list: K_a, K_d, Hill coefficient, number of binding sites per molecule of a binding agent.

1. Ligand concentration necessary to obtain $\bar{v} = 0.5$ in the rectangular hyperbola ______________________.
2. The ligand concentration corresponding to midpoint of the sigmoid curve ______________________.
3. The slope of the plot, log [P]/[PL] versus Log [L] ______________________.
4. The slope divided by the intercept of the double reciprocal plot ______________________.
5. The slope of the Scatchard plot, with its sign changed from negative to positive ______________________
6. The intercept on the abscissa of the Scatchard plot ______________________.

(*Items 150-157.*)

195

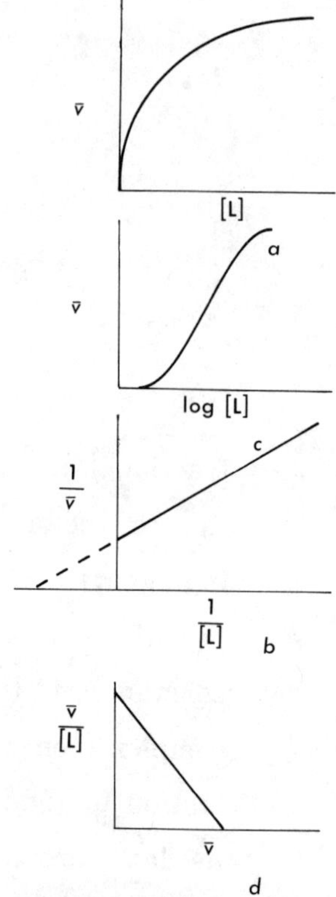

196

1. K_d
2. K_d
3. Hill coefficient
4. K_d
5. K_a
6. number of binding sites per molecule

Appendix A

What follows is a concise recapitulation of the derivation of the Michaelis-Menten equation under the steady-state approximation.

$$E + S \underset{k_{-1}}{\overset{k_1}{\rightleftharpoons}} ES \xrightarrow{k_2} P + E$$

$$\frac{d[ES]}{dt} = k_1[E][S] - k_{-1}[ES] - k_2[ES]$$

Since $[E] = [E_t] - [ES]$

$$\frac{d[ES]}{dt} = k_1([E_t] - [ES])[S] - (k_2 + k_{-1})[ES]$$

By the steady-state approximation

$$\frac{d[ES]}{dt} = 0$$

Therefore

$$k_1([E_t] - [ES])[S] - (k_2 + k_{-1})[ES] = 0$$

Expanding

$$k_1[E_t][S] = k_1[ES][S] + k_2[ES] + k_{-1}[ES]$$

$$k_1[E_t][S] = (k_1[S] + k_2 + k_{-1})[ES]$$

Solving for $[ES]$

$$[ES] = \frac{k_1[E_t][S]}{k_1[S] + k_2 + k_{-1}}$$

Dividing both the numerator and the denominator of the fraction by k_1

$$[ES] = \frac{[E_t][S]}{[S] + \dfrac{k_2 + k_{-1}}{k_1}}$$

According to the basic rate equation for the enzymatic process, $v = k_2[ES]$. By introducing our value above for $[ES]$ into the basic rate equation we obtain

$$v = \frac{k_2[E_t][S]}{[S] + \dfrac{k_2 + k_{-1}}{k_1}}$$

If we define K_m as

$$\frac{k_2 + k_{-1}}{k_1}$$

and define V_{max} as $k_2[E_t]$, we obtain:

$$v = \frac{V_{max} \cdot [S]}{K_m + [S]}$$

(The quantities that disàppear from the derivation under the equilibrium assumption are indicated by the shadowed background.)

Appendix B

Here are demonstrations of the three linear transformations of the Michaelis-Menten equation. The purpose in all three cases is to recast the equation into the general form, $y = a + b \cdot x$ where x and y are the variables, a is the intercept, and b the slope of a straight-line graph.

(1) The method of Lineweaver and Burk (due originally to Woolf):

$$v = \frac{V_{max} \cdot [S]}{K_m + [S]}$$

inverting

$$\frac{1}{v} = \frac{K_m + [S]}{V_{max} \cdot [S]} = \frac{K_m}{V_{max} \cdot [S]} + \frac{[S]}{V_{max} \cdot [S]}$$

$$\frac{1}{v} = \frac{1}{V_{max}} + \frac{K_m}{V_{max}} \cdot \frac{1}{[S]}$$

$$y = a + b \cdot x$$

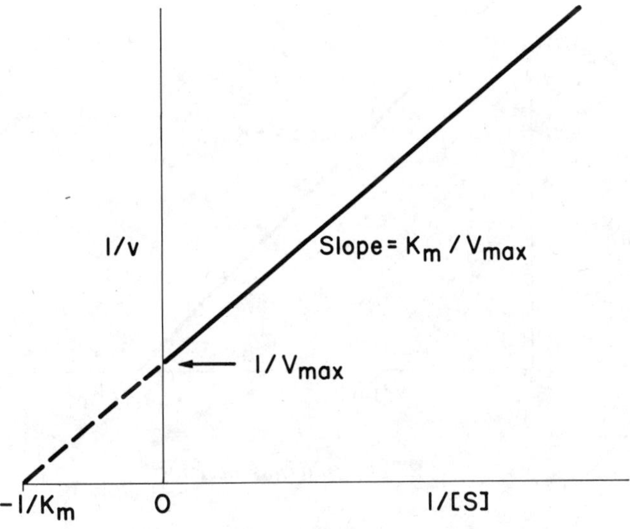

1. Lineweaver-Burk Plot

(2) The method of Augustinsson, often attributed to Hofstee:

$$v = \frac{V_{max} \cdot [S]}{K_m + [S]}$$

Dividing the numerator and denominator of the fraction on the right-hand side by $[S]$

$$v = \frac{V_{max}}{\frac{K_m}{[S]} + 1}$$

Therefore

$$v\left(\frac{K_m}{[S]} + 1\right) = V_{max}$$

and

$$\frac{v}{[S]} \cdot K_m + v = V_{max}$$

or

$$v = V_{max} - K_m \cdot \frac{v}{[S]}$$

$$y = a + b \cdot x$$

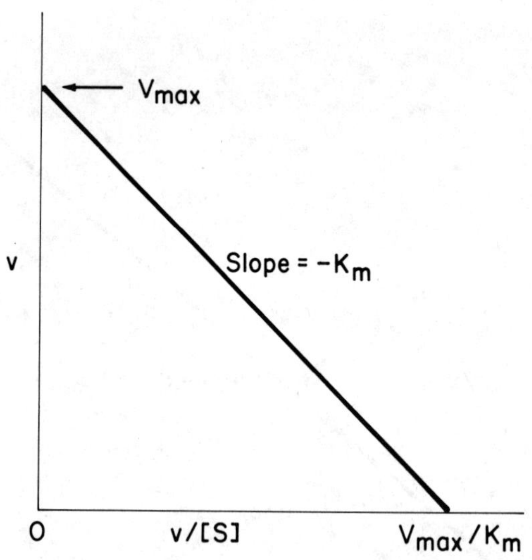

2. Augustinsson Plot

(3) The method of Woolf. Take the equation used in the method of Lineweaver and Burk:

$$\frac{1}{v} = \frac{1}{V_{max}} + \frac{K_m}{V_{max}} \cdot \frac{1}{[S]}$$

Multiply through by $[S]$ to obtain

$$\frac{[S]}{v} = \frac{[S]}{V_{max}} + \frac{K_m}{V_{max}}$$

or

$$\frac{[S]}{v} = \frac{K_m}{V_{max}} + \frac{1}{V_{max}} \cdot [S]$$

$$y = a + b \cdot x$$

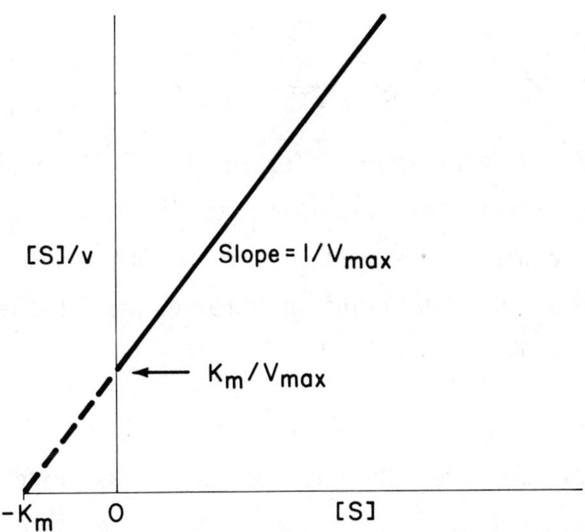

3. Woolf Plot. (Note that Woolf originally proposed all three methods.)

Appendix C

Here we will prove that the graphic method of Dixon yields the value of K_i. The equation for competitive inhibition:

$$v = \frac{V_{max}}{1 + \frac{K_m}{[S]}\left(1 + \frac{[I]}{K_i}\right)}$$

is written in a reciprocal form

$$\frac{1}{v} = \frac{K_m}{V_{max}[S]} + \frac{1}{V_{max}} + \frac{K_m}{V_{max}[S]} \cdot \frac{[I]}{K_i}$$

In the Dixon method the velocity is determined at various inhibitor concentrations at a constant substrate concentration, yielding a straight line plot of $1/v$ against $[I]$. The procedure is repeated at a second substrate concentration, yielding a second straight line. The point of intersection of these lines (Item 121) gives $-K_i$ directly. That is, at this point $[I] = -K_i$.

Proof. At the point of intersection the values $1/v$ and $[I]$ will be the same for the two lines; V_{max} is also the same since the inhibition is competitive. Therefore

$$\frac{K_m}{V_{max}[S_1]} + \frac{1}{V_{max}} + \frac{K_m}{V_{max}[S_1]} \cdot \frac{[I]}{K_i} = \frac{K_m}{V_{max}[S_2]} + \frac{1}{V_{max}} + \frac{K_m}{V_{max}[S_2]} \cdot \frac{[I]}{K_i}$$

Multiplying through by V_{max}

$$\frac{K_m}{[S_1]} + 1 + \frac{K_m}{[S_1]} \cdot \frac{[I]}{K_i} = \frac{K_m}{[S_2]} + 1 + \frac{K_m}{[S_2]} \cdot \frac{[I]}{K_i}$$

Subtract 1 from each side, and collect terms in which $\frac{K_m}{[S]}$ appears

$$\frac{K_m}{[S_1]}\left(1 + \frac{[I]}{K_i}\right) = \frac{K_m}{[S_2]}\left(1 + \frac{[I]}{K_i}\right)$$

Now for this equation to be true, either $[S_1]$ must equal $[S_2]$ (which is not the case), or the quantity $\left(1 + \frac{[I]}{K_i}\right)$ must be zero. The latter requires that $[I] = -K_i$.

Appendix D

What follows is a reprint of the paper of Briggs and Haldane, from the *Biochemical Journal*, *19*:338-339, 1925.*

L. A NOTE ON THE KINETICS OF ENZYME ACTION.

BY GEORGE EDWARD BRIGGS
AND JOHN BURDON SANDERSON HALDANE.
(*From the Botanical and Biochemical Laboratories, Cambridge.*)

(*Received March 9th, 1925.*)

THE equation of Michaelis and Menten [1913] has been applied with success by Kuhn [1924] and others to numerous cases of enzyme action. It is therefore desirable to examine its theoretical basis. Consider the irreversible reaction $A \rightarrow B$, unimolecular as regards A, and catalysed by an enzyme. Suppose one molecule of A to combine reversibly with one of enzyme, the compound then changing irreversibly into free enzyme and B, where B may represent several molecules. We may represent this as:

$$\underset{(a-x)\,(e-p)}{A+E} \rightleftharpoons \underset{p}{AE} \rightarrow \underset{x}{B} + E.$$

Now let a be the initial concentration of A, e the total concentration of enzyme, x the concentration of B produced after time t, and p the concentration of enzyme combined with substrate at time t. We suppose e and p to be negligibly small compared with a and x. Then by the laws of mass action

$$\frac{dp}{dt} = k_1 (a-x)(e-p) - k_2 p - k_3 p,$$

*Reproduced by permission. In a letter dated 18 August 1965 consenting to our reproduction of this paper, Professor Briggs wrote as follows on the origin of his joint authorship with Haldane:

> When I came back and started research after World War I, I was interested in the kinetics of photosynthesis and enzyme activity, and, as has been my wont, I evolved the simple formulation from elementary first principles. Later I discovered the papers of Michaelis and Menten, and Van Slyke and Cullen. Some years later Haldane in talking to F.F. Blackman said he had a new idea on enzyme kinetics. Having heard it Blackman said "Briggs has been teaching that for some time now" to which Haldane suggested that we ought to write a joint paper — we did.

where k_1, k_2, k_3 are the velocity constants of the reactions

$$A + E \rightarrow AE, \quad AE \rightarrow A + E, \quad \text{and} \quad AE \rightarrow B + E,$$

respectively. Now since p is always negligible compared with x and $a - x$, its rate of change must, except during the first instant of the reaction, be negligible compared with theirs.

For during the remainder of the reaction p diminishes from a value not exceeding e to zero, whilst x increases from zero to a. Thus the average value of $-\frac{dp}{dt} \div \frac{dx}{dt}$ is less than $\frac{e}{a}$. And provided $\frac{e}{a}$ is small it is clear that if the amount of combined enzyme decreased for a measurable time at a rate comparable with that of its substrate the reaction would come to an end. To take a concrete example Kuhn [1924] calculates that a yeast saccharase molecule at 15° and p_{H} 4·6 can invert 100 or more molecules of sucrose per second. Even if the enzyme concentration is so unusually large that the inversion of a strong sucrose solution is half completed in 10 minutes, $\frac{a}{e}$ cannot be less than 120,000, and if $-\frac{dp}{dt}$ attained 1 % of the value of $\frac{dx}{dt}$ for 1 second the reaction would stop owing to all the enzyme being set free. (Actually it may be shown that

$$\frac{dp}{dt} = \frac{-k_3 (k_2 + k_3)\, e\, (a - x)}{k_1 \left(a - x + \frac{k_2 + k_3}{k_1}\right)^3}.)$$

Hence

$$k_1 (a - x)(e - p) - k_2 p - k_3 p = 0.$$

$$\therefore\ p = \frac{k_1 e (a - x)}{k_2 + k_3 + k_1 (a - x)}$$

$$= \frac{e (a - x)}{a - x + \frac{k_2 + k_3}{k_1}}$$

$$\therefore\ \frac{dx}{dt} = k_3 p = \frac{k_3 e (a - x)}{a - x + \frac{k_2 + k_3}{k_1}}.$$

This is Michaelis and Menten's [1913] equation, $(k_2 + k_3)/k_1$ representing their constant K_s. They assume that the reaction

$$A + E \rightleftharpoons AE$$

is always practically in equilibrium, and K_s its equilibrium constant, *i.e.* that k_3 is negligible in comparison with k_2. Van Slyke and Cullen [1914] on the other hand assumed that the first stage of the reaction was irreversible, *i.e.* $k_2 = 0$, and arrived at the same equation. It is clear, however, that data as to the course of a reaction can give no indication of the ratio of k_2 and k_3, though when the velocity of the observed reaction, and hence k_3, is very large, the upper limit to k_2 deducible from the kinetic theory may possibly prove to be of the same order of magnitude.

It may be remarked that with this modification of their theory, Michaelis and Menten's analysis of the effects of the products of the reaction, or other substances which combine with the enzyme, still holds good.

REFERENCES.

Kuhn (1924). Oppenheimer's "Die Fermente und ihre Wirkungen," 185 et seq.
Michaelis and Menten (1913). *Biochem. Z.* **49**, 333.
Van Slyke and Cullen (1914). *J. Biol. Chem.* **19**, 141.

Examination Problems on Enzyme Kinetics

SECTION A. Arithmetic and Graphic Problems.

Problem 1. Given, that an enzymatic reaction shows the following parameters:

$V_{max} = 1000$ micromoles of substrate consumed/min · mg protein

$K_m = 5 \times 10^{-4}\ M$

Calculate v for the following concentrations of substrate:

(*a*) $2 \times 10^{-4}\ M$

(*b*) $5 \times 10^{-4}\ M$

(*c*) $1 \times 10^{-3}\ M$

(*d*) $5 \times 10^{-3}\ M$

(*e*) $1 \times 10^{-2}\ M$

Problem 2. Similarly, given for another reaction that $V_{max} = 50{,}000\ \text{min}^{-1}$ and $K_m = 1 \times 10^{-6}\ M$, calculate v at:

(*a*) $5 \times 10^{-8}\ M$

(*b*) $1 \times 10^{-7}\ M$

(*c*) $8 \times 10^{-7}\ M$

(*d*) $3 \times 10^{-6}\ M$

(*e*) $2 \times 10^{-4}\ M$

Problem 3. In the two preceding problems identify concentration ranges where the velocity approximates, within 10 per cent, (*a*) first-order kinetics or (*b*) zero order kinetics (rate independent of substrate concentration).

Problems 4 to 6. The following observations describe the reciprocals of the velocity of three enzymatic reactions.

$[S]$ mM	$\frac{1}{[S]}$ M^{-1}	$\frac{1}{v_4}$ min	$\frac{1}{v_5}$ min	$\frac{1}{v_6}$ min
1.0	1000	0.014	0.024	0.028
0.5	2000	0.018	0.028	0.035
0.25	4000	0.025	0.035	0.050
0.167	6000	0.033	0.043	0.065
0.125	8000	0.040	0.050	0.080

Take a sheet of graph paper, and determine by plotting on a single pair of coordinate axes whether these reactions correspond to the Michaelis-Menten equation. If so, determine from the plots the values of K_m and V_{max} in each case.

Problem 7. In these particular cases, why do the rates v_4 and v_5 approach each other at low substrate levels? Why do the rates v_5 and v_6 approach each other at high substrate levels?

SECTION B. Objective Problems. Each of the sets of lettered subjects below is followed by a numbered list of phrases or statements. For each numbered phrase or statement insert the letter:

A, If the phrase or statement is associated with (*A*) only

B, If the phrase or statement is associated with (*B*) only

C, If the phrase or statement is associated with both (*A*) and (*B*)

D, If the phrase or statement is associated with neither (*A*) nor (*B*)

First Set: For an enzymatic reaction corresponding to the formulation

$$E + S \underset{k_{-1}}{\overset{k_1}{\rightleftharpoons}} ES \xrightarrow{k_2} E + P$$

A. K_m

B. V_{max}

C. Both

D. Neither

( ) 1. Measures the affinity of the substrate for the enzyme.

( ) 2. Modified by a suitable inhibitor.

( ) 3. If expressed in suitable terms, becomes identical with the *turnover number*.

( ) 4. In the accompanying plot of rate versus substrate concentration, represented approximately by mark No. 1.

( ) 5. Represented approximately by mark No. 2.

( ) 6. In the second plot, this parameter is measured (directly or reciprocally) by the interval marked 4.

( ) 7. This parameter is measured (directly or reciprocally) by the interval marked 3.

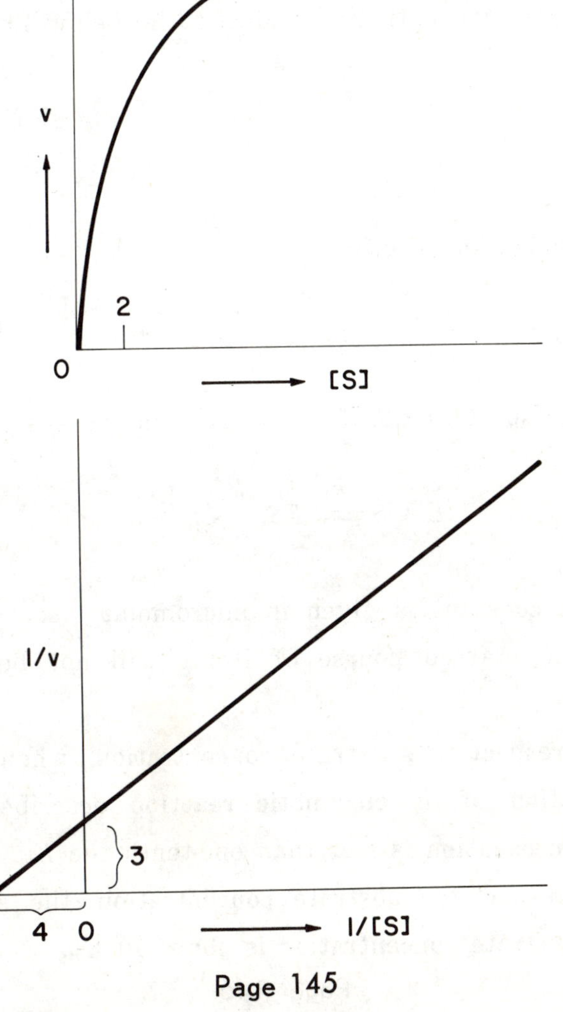

( ) 8. Equal to $k_2[ES]$.

( ) 9. This parameter is most conservatively taken to represent $\frac{k_{-1} + k_2}{k_1}$, although in some cases it may prove to be equal to k_{-1}/k_1.

( ) 10. If the rate of an enzymatic reaction is as tabulated below, the value of this parameter, by inspection, is very near the value given in line (*a*) of the accompanying table:

	$[S]$, mM	v, *mmoles/min*
(*a*)	1	405
(*b*)	2	533
(*c*)	50	800

( ) 11. Changed by the presence of a competitive inhibitor.

( ) 12. Changed by the presence of a non-competitive inhibitor.

( ) 13. Needed to calculate the rate of a given enzyme reaction at a given substrate concentration known to lie below 10 K_m.

Second Set:

A. A first-order reaction

B. A second-order reaction

C. Both

D. Neither

( ) 14. With respect to *ES*, if $v = k_2[ES]$, in the system

$$E + S \underset{k_{-1}}{\overset{k_1}{\rightleftharpoons}} ES \xrightarrow{k_2} E + P$$

( ) 15. A rate constant is given in micromolar^{-1} sec^{-1}. (Students following the starred course of Items will not be prepared for this question.)

( ) 16. With respect to substrate concentration, when the substrate concentration of an enzymatic reaction described by the Michaelis-Menten equation is less than one-tenth the K_m.

( ) 17. With respect to substrate concentration, the same reaction, when the substrate concentration is above 10 K_m.

Third Set: With reference to the system

$$S + E \underset{k_{-1}}{\overset{k_1}{\rightleftharpoons}} ES \xrightarrow{k_2} E + P$$

A. Will be true where the equilibrium assumption of Michaelis and Menten applies

B. Will be true where the steady-state approximation of Briggs and Haldane applies

C. Will be true if either applies

D. Will not be true if either applies

( ) 18. A plot of velocity versus substrate concentration takes the form of a rectangular hyperbola.

( ) 19. $v = k_2[ES]$

( ) 20. $K_m = k_{-1}/k_1$

( ) 21. The *turnover number* equals k_2.

( ) 22. $d[ES]/dt = 0$

( ) 23. $k_1[E][S] = k_{-1}[ES]$

( ) 24. $k_2[E_t] = V_{max}$

( ) 25. $v = \frac{V_{max} + [S]}{K_m \cdot [S]}$

Fourth Set: In an enzymatic reaction that follows the Michaelis-Menten equation

A. Competitive inhibition

B. Non-competitive inhibition

C. Both

D. Neither

( ) 26. Requires that the inhibitor form a complex with the enzyme.

( ) 27. Established, if the Lineweaver-Burk plot in the presence of an inhibitor proves to be a straight line.

( ) 28. In the presence of an inhibitor, the Lineweaver-Burk plot will be a straight line with a different slope but with the same intercept on the ordinate as that obtained in the absence of the inhibitor.

( ) 29. The presence of the inhibitor decreases the V_{max}.

( ) 30. The presence of the inhibitor decreases the K_m.

( ) 31. The inhibitor must be an analog that can itself undergo the typical enzymatic reaction.

( ) 32. The greater the slope of the Lineweaver-Burk plot obtained under this action, the more potent the inhibition.

( ) 33. Requires that the inhibitor react with the enzyme at some point other than the reactive site.

Fifth Set:

A. pH
B. Temperature
C. Both
D. Neither

( ) 34. The accompanying curve might well illustrate the effect of this variable on the velocity of an enzymatic reaction.

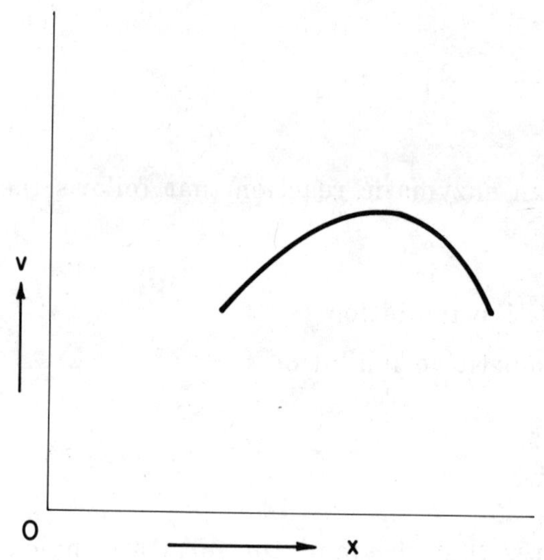

( ) 35. The decrease in enzyme activity as this variable increases may well in a given instance arise principally from an effect on the substrate.

( ) 36. An increase in this variable might well decrease enzymatic reactivity by denaturing the enzyme.

SECTION C. Provocative problems, which are not necessarily answered by the material that has gone before.

Problem 1. Recently an investigator reported that the K_m for transport of a sugar out of a given bacterial cell is about 100 times the K_m for its inward transport. The initial rate of transfer was observed in each direction, and the inward and outward transfers each followed the formulation

$$E + S \underset{k_{-1}}{\overset{k_1}{\rightleftharpoons}} ES \overset{k_2}{\longrightarrow} E + P$$

but with a hundred-fold difference in the K_m values. That difference may be understood to permit the solute to reach as much as 100 times as high a concentration inside the cell as outside.

The investigator further showed that adding 2, 4-dinitrophenol brought the K_m for exodus almost completely down to that for entry, so that the solute was no longer appreciably concentrated.

The following two major possible explanations of uphill transport have been considered:

(*A*) That the binding site, once it receives the solute, is driven inward by a vectorial force, to cause the solute to be transported inward against the gradient.

(*B*) That energy is not applied directly to the translocation, but instead the nature of the reactive site is changed at the inward surface of the membrane in such a way that its affinity for the solute is decreased, so that the solute can be released at a concentration in the inner phase higher than that in the outer.

Does the kinetic evidence cited permit a choice in favor of the latter explanation? ______

Problem 2. An investigator observes that when two analogs compete for a given enzymatic reaction in which they both react typically, each substrate shows a K_i value in inhibiting the reaction of the other, just equal to its own K_m for the reaction. He concludes therefore that the K_m's are dissociation constants for the enzyme-substrate, since a textbook states that $K_i = \frac{k_{-1}}{k_1}$ for the complex, *EI*. Do you agree? ______

Problem 3. If glucose and mannose were present in equal concentrations in the reaction mixture with the enzyme of Item 100, and if one used an analytical method that measured in moles the *total reaction due to both sugars,* what values of K_m and V_{max} would be obtained? Is there any aspect of this case as described in Item 100 that would encourage one to suppose that the two K_m values actually measure the relative affinities of glucose and mannose for the enzyme? ______

Problem 4. Many enzyme reactions are much more complicated than those we have been considering. For instance, the dehydrogenases catalyze reactions of the type

$$AH_2 + \begin{matrix} oxidized \\ pyridine \\ nucleotide \end{matrix} \overset{enzyme}{\rightleftharpoons} A + \begin{matrix} reduced \\ pyridine \\ nucleotide \end{matrix}$$

where AH_2 and A refer to the reduced and oxidized forms of the substrate, respectively. These might be ethanol and acetaldehyde (for alcohol dehydrogenase); or lactate and pyruvate (for lactate dehydrogenase).

These enzymes frequently operate through the formation of a ternary complex of the type

$$(Enzyme)\ (AH_2)\ (oxidized\ pyridine\ nucleotide)$$

Suppose that by plotting the reciprocal of the velocity, $1/v$, against $1/[AH_2]$ at a fixed concentration of the oxidized pyridine nucleotide, one gets a straight line corresponding to a given value of K_m. Suppose, conversely, one also gets a straight line and another K_m value by plotting $1/v$ against 1/[oxid. pyr. nucleotide] at a fixed concentration of the reduced substrate. Should a special theoretical significance be attributed to these K_m values, i.e., is either a dissociation constant? ______

Problem 5. Suppose one has a catalytic process (either a chemical reaction or a transport) for which two analogs compete on an even basis; i.e., the two happen to have the same K_m (and K_i) values. Suppose further that the first of these substrates is introduced into the system in ^{14}C-labeled form to provide a certain number of counts per min per ml. Now let us add to the system a given molar concentration of either (*a*) additional unlabeled substrate I, or (*b*) the analogous substrate II. The figure on the next page shows the slowing of the entry of ^{14}C into the process in the presence of II.

Will the entry of ^{14}C into the process be slowed any more by the inhibitory analog than it is slowed by the excess of the substrate itself? ______ (Note the paradox, emphasized by the Augustinsson plot, that we use the term *maximal velocity* for the conditions under which the velocity, *relative to concentration* is | minimal | | maximal |.)

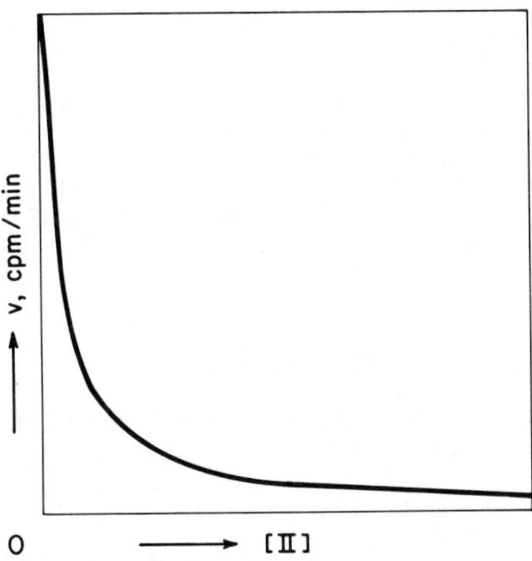

Does the reactive site in this case recognize any difference between the substrate and its analog? ______ Could one determine the K_i of the substrate by plotting its effectiveness at various concentrations in suppressing the uptake of the original amount of its ^{14}C-labeled form, i.e., by its inhibitory action? ______ Is this effect what we ordinarily mean by the term *substrate inhibition* (*Item 99*)? ______

Problem 6. A number of transport mediators are known to have rather broad tolerances in their specificities, e.g., for whole classes of sugars or amino acids. In some of these cases V_{max} tends to have about the same value for all members of the class. If that is found to be the case, could an inhibition analysis serve to explore the structure of the active site, by seeing what structural features of the substrate favor the affinity? ______

Problem 7. The V_{max} for the transfer of a substance across the renal tubule, expressed as the quantity of the substance that can be transferred per organism per minute, is known as T_m, the maximal tubular transfer capacity. For glucose in man, T_m averages about 375 mg per minute. The volume of glomerular fluid formed by the ultrafiltration of plasma in man averages about 125 ml per minute. Any quantity of glucose not resorbed during the passage of this fluid through the proximal tubule will of course be lost into the urine.

To provide the quantity of glucose that can be resorbed, the glomerular filtrate would need to contain 375/125 or 3 mg of glucose per ml (300 mg per 100 ml). From the data given, can you conclude that the urinary "threshold" concentration of plasma glucose (i.e., the minimal concentration needed to induce glucosuria) will lie at about 300 mg per 100 ml? ______ Considering only what you have learned in this program, can you calculate the urinary threshold from the data given? ______ Would you expect the threshold to represent a precise concentration above which all additional glucose is excreted? ______

Problem 8. If we take the more general case of an enzyme reaction where enough *product* is present so that the backward reaction cannot be ignored

$$E + S \underset{k_{-1}}{\overset{k_1}{\rightleftharpoons}} ES \underset{k_{-2}}{\overset{k_2}{\rightleftharpoons}} E + P$$

one can derive under the steady-state approximation a more general rate equation beginning as follows (note that the underlined term is new):

$$v = \frac{dP}{dt} = k_2 [ES] - \underline{k_{-2} [E] [P]}$$

$$\frac{d[ES]}{dt} = k_1 [E][S] + \underline{k_{-2} [E] [P]} - (k_{-1} + k_2) [ES] = 0$$

(The steady-state approximation)

One then goes through the first three steps used in the other derivations (see Item 65); and by finding the lowest common denominator arranges the resultant expression for v so that it has the form of a single fraction

$$v = \frac{(k_1 k_2 [S] - k_{-1} k_{-2} [P]) [E_t]}{k_1 [S] + k_{-2} [P] + k_{-1} + k_2}$$

This is the general rate equation for the above reaction. Show that when P is taken to be zero, this general equation simplifies to the Michaelis-Menten equation.

Problem 9. If one plots the amount of MbO_2 formed as a function of pressure of oxygen, for the reaction of myoglobin with oxygen

$$Mb + O_2 \rightleftharpoons MbO_2$$

he obtains a rectangular hyperbola. He also obtains a rectangular hyperbola if he plots the amount of *HA* formed as a function of $[H^+]$ in the reaction

$$A^- + H^+ \rightleftharpoons HA$$

Similarly, the amount of a solute an adsorbent takes up can be described as a function of solute concentration by a curve of the same form, according to the Langmuir adsoption isotherm. How do these cases differ from the kinetics of enzyme reactions or transport described in this program?

The last parallelism in the above list may suggest to you that we could plot the rate of an enzyme reaction as a function of the negative logarithm of the substrate concentration, just as we plot A^- formation as a function of the negative log of $[H^+]$. Which of the following descriptions would then apply for a catalytic reaction following Michaelis-Menten kinetics?

a. The curves for a series of analogous substrates would all have a common sigmoid shape, but their lateral positions would differ according to the values of K_m.

b. The curves would have different slopes but their midpoints would have identical positions.

(We may perhaps regard it as a historical fortuity that the logarithmic plot is preferred for the titration of weak acids and bases, whereas linear transformations are preferred for plotting the kinetics of catalytic reactions.)

(The common ground between enzyme kinetics and binding studies may be brought out once more by noting that the data plotted in Item 130, to illustrate sigmoidal rather than hyperbolic kinetics, are actually the classical data for the per cent saturation of hemoglobin as a function of the oxygen pressure!)

Problem 10. The concentration of protein present in the dialysis sac during a series of equilibrium dialysis experiments was $10^{-4}M$. In these experiments the following concentrations of total ligand in the dialysis fluid and the sac were observed at equilibrium. Complete the table:

Total [ligand] in sac $[L_t]$	Total [ligand] in dialysis fluid, $[L_f]$	$[L_t] - [L_f] =$ [Bound ligand]	[Bound ligand] / [Protein]
M	M	M	
0.209×10^{-4}	0.200×10^{-4}	0.009×10^{-4}	0.009
0.623	0.600		
1.333	1.000		
2.500	2.000		
4.667	4.000		
10.88	10.00		
20.91	20.00		
100.98	100.00		

Now complete the following graph, first by plotting the data on the coordinates as labeled at the left and the bottom; then calculate the reciprocals of the data and plot according to the coordinate labels at the right and the top.

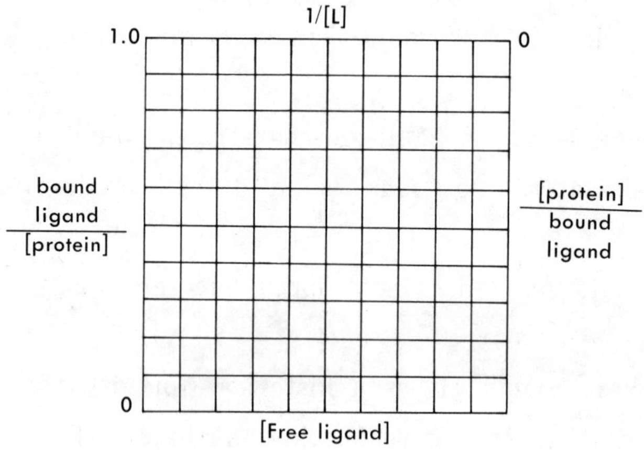

1. How many moles of ligand are bound per mole of protein?
2. What is the equilibrium constant for the reaction? Is it an association or dissociation constant?

Answers

SECTION A. Answers to Arithmetic and Graphic Problems:

Problem 1. 286; 500; 670; 910; 950

Problem 2. 2400; 4550; 22,000; 37,500; 49,800

Problem 3. First-order, in Problem 2, 5×10^{-8} and $1 \times 10^{-7}\ M$
Zero-order, in Problem 1, 5×10^{-3} and $1 \times 10^{-2}\ M$;
in Problem 2, $2 \times 10^{-4} M$

Problem 4. $V_{max} = 100$; $K_m = 3.8 \times 10^{-4}\ M$

Problem 5. $V_{max} = 50$; $K_m = 1.9 \times 10^{-4}\ M$

Problem 6. $V_{max} = 50$; $K_m = 3.75 \times 10^{-4}\ M$

Problem 7. For Problems 4 and 5, $\frac{V_{max}}{K_m}$ has identical values; hence the reaction rates approach the same values at low substrate concentrations.
For Problems 5 and 6, the V_{max} values are identical; hence the rates approach each other at high substrate levels.

SECTION B. Answers to Objective Problems:

(D) 1.
(C) 2.
(B) 3.
(B) 4.
(A) 5.
(A) 6.
(B) 7.
(D) 8.
(A) 9.
(A) 10.
(A) 11.
(B) 12.
(C) 13.
(A) 14.
(B) 15.
(A) 16.
(D) 17.
(C) 18.
(C) 19.
(A) 20.
(C) 21.
(C) 22.
(A) 23.
(C) 24.
(D) 25.
(C) 26.
(D) 27.
(A) 28.
(B) 29.
(D) 30.
(D) 31.
(C) 32.
(D) 33.
(C) 34.
(A) 35.
(C) 36.

SECTION C. Answers to Provocative Problems:

Problem 1. No. To draw an affirmative conclusion would be to make the equilibrium assumption, i.e., to assume that the K_m values measure the affinity of the solute for the transport site at each surface of the membrane. The change in K_m noted could instead result from effects of dinitrophenol on process k_2, i.e., on the presumed vectorial event.

Problem 2. No. Because most studies of inhibitors have been made with analogs unable to serve as typical substrates, the special case

$$K_i = \frac{k_{-1}}{k_1}$$

is often cited as if it were a universal definition for K_i.

Problem 3. A value for K_m intermediate between those for glucose and mannose would be expected, namely $7.5 \times 10^{-5}\,M$. The V_{max} in this case would be the same as it is for either sugar alone. The fact that V_{max} was found to be the same for the two substrates indicates that k_2 was also the same. Hence all the difference in the K_m values must appear in the values for $\frac{k_{-1}}{k_1}$.

Problem 4. No. These numbers are empirical; they are not even characteristic constants of the system, in the sense that K_m values otherwise may be.

Problem 5. No; minimal (in contrast, the highest value of $\frac{v}{[S]}$, namely

$$\frac{V_{max}}{K_m}$$

is approached only at very low concentrations); no; yes; no.

Problem 6. Yes. The constancy of V_{max} probably means that k_2 is invariant. Hence K_m or K_i should actually measure affinity.

Problem 7. All three questions take *no* answers on theoretical grounds, but in fact one would not be entirely right to respond thus. It is true, one cannot calculate the rate of a catalytic process at a given substrate concentration unless *both* the V_{max} and the K_m are given. A sufficient concentration of glucose in the fluid in the tubular lumen will be required to cause the maximal velocity to be approximated, which means that as the blood sugar rises, theoretically one should not pass abruptly from a situation at which *no* sugar at all appears in the urine, to one in which all the additional plasma sugar is excreted. In actual fact, the K_m of glucose for the transport is so low, however, that a fairly definite threshold is observed at about 300 mg per 100 ml.

Problem 8. If we study the *initial rate* so that $[P]$ can be taken as zero, all terms containing $[P]$ drop out of the general equation, thus:

$$v = \frac{k_1 k_2 [S] \cdot [E_t]}{k_1 [S] + k_{-1} + k_2}$$

Dividing both numerator and denominator of the fraction by k_1,

$$v = \frac{k_2 [S] [E_t]}{[S] + \frac{k_{-1} + k_2}{k_1}}$$

When V_{max} is introduced in place of $k_2 [E_t]$, and K_m for $\frac{k_{-1} + k_2}{k_1}$ this equation becomes the Michaelis-Menten equation.

Problem 9. In the case of the formation of MbO_2, of HA or of adsorption complexes, the reaction comes to equilibrium because no further chemical reaction can presumably take place. In the kinetics of catalysis we have seen that the formation of ES may reach an equilibrium with respect to E and S because the subsequent catalytic event is relatively slow. The concentration of ES will in any event remain nearly constant for an interval of time, because its breakdown both to $E + S$ and to $E + P$ will approximate its rate of formation. For the non-catalytic case we measure *directly* the formation of the complex corresponding to ES. For the catalytic case we measure *indirectly* the formation of ES by observing v, which of course is directly proportional to $[ES]$. (The applicability of the adsorption isotherm shows that the adsorbent has a limited number of binding sites for the solute.)

A plot of v versus $-\log [S]$ (i.e., $p[S]$) for catalytic reactions should give similar sigmoid curves for a group of analogous, reactive substrates, the position of each curve varying with the value of pK_m, just as titration curves take positions depending on the value of pK' for the weak acid. (In each case we may normalize the curves, by plotting *percent dissociation* versus pH for the dissociation of weak acids and by plotting *percent of maximal velocity* versus $p[S]$ for catalysis.)

Problem 10. Either plot of the data shows that one mole of ligand is bound per mole of protein, and that the equilibrium constant is 2×10^{-4}M, since the protein is half-saturated at that concentration of free ligand. This is a dissociation constant.

$[L_f]$	$[L_t] - [L_f]$	$\frac{\text{[Bound ligand]}}{\text{[protein]}}$
M × 10^4	M × 10^4	
0.200	0.009	0.009
0.600	0.023	0.023
1.000	0.333	0.333
2.000	0.500	0.500
4.000	0.667	0.667
10.00	0.88	0.88
20.00	0.91	0.91
100.00	0.98	0.98

Recommended Supplementary References

CHEMICAL KINETICS

1. E.L. King: *How Chemical Reactions Occur: An Introduction to Chemical Kinetics and Reaction Mechanisms.* New York, W.A. Benjamin, Inc., 1963. A very readable introduction to the topic.
2. A.A. Frost and R.G. Pearson: *Kinetics and Mechanisms: A Study of Homogeneous Chemical Reactions.* Second edition. New York, John Wiley & Sons, Inc., 1961. A clear and comprehensive treatment which is highly recommended.
3. S.L. Freiss, E.S. Lewis, and A. Weissberger (Eds.): *Investigation of Rates and Mechanisms of Reactions.* In, A. Weissberger (Ed.): *Techniques of Organic Chemistry.* Volume VIII (in two parts). New York, Interscience Publishers, Inc. Part I, second edition, 1961. Part II, second edition, 1963. Found on the shelves of many practicing kineticists.

ENZYME KINETICS

4. M. Dixon and E.C. Webb: *Enzymes.* Second edition. New York, Academic Press, 1964. Probably the best single book on the properties of enzymes, with extensive discussions of enzyme kinetics.
5. J.M. Reiner: *Behavior of Enzyme Systems: An Analysis of Kinetics and Mechanisms.* Minneapolis, Burgess Publishing Co., 1959. A detailed outline of the kinetics of enzyme reactions which makes few assumptions about the mathematical ability of its readers. Highly recommended for students who wish to acquire some proficiency in this area.
6. J.B.S. Haldane: *Enzymes.* Reprint. Cambridge, Massachusetts, The M.I.T. Press, 1965. A classic, providing an interesting account of the status of enzymology in 1930. References cited.
7. C.F. Walter and M.F. Morales: An analogue computer investigation of certain issues in enzyme kinetics. *J. Biol. Chem., 239:*1277-1283, 1964.
8. J.E. Dowd and D.S. Riggs: A comparison of estimates of Michaelis Menten kinetic constants from various linear transformations. *J. Biol. Chem., 240:*863-869, 1965. (See also G.N. Wilkinson: Statistical estimations in enzyme kinetics. *Biochem. J., 80:*324-332, 1961.)
9. L.N. Johnson and D.C. Phillips: Structure of some crystalline lysozyme inhibitor complexes determined by x-ray analysis at 6 Å resolution. *Nature, 206:*761-763, 1965; and C.C.F. Blake, et al.: Structure of hen egg-white lysozyme. *Nature, 206:*757-761, 1965.
10. J. Monod, J. Wyman, and J.P. Changeux: On the nature of allosteric transitions: a plausible model. *J. Mol. Biol., 12:*88-118, 1965.
11. O.H. Lowry and J.V. Passonneau: *A Flexible System of Enzymatic Analysis.* New York, Academic Press, 1972. Provides valuable guides for the selection of conditions for enzyme assays, along with description of methods, including use of "recycling" to increase sensitivity.

Index

All references in this index are to *page* number rather than item number.